T0139868

Gaming Media and Social Effects

Editor-in-chief

Henry Been-Lirn Duh, University of Tasmania, Hobart, TAS, Australia

Series editor

Anton Nijholt, University of Twente, Enschede, The Netherlands

The scope of this book series is inter-disciplinary and it covers the technical aspect of gaming (software and hardware) and its social effects (sociological and psychological). This book series serves as a quick platform for publishing top-quality books on emerging or hot topics in gaming and its social effects. The series is also targeted at different levels of exposition, ranging from introductory tutorial to advanced research topics, depending on the objectives of the authors.

More information about this series at http://www.springer.com/series/11864

Daniel Cermak-Sassenrath
Editor

Playful Disruption of Digital Media

 Springer

Editor
Daniel Cermak-Sassenrath
Center for Computer Games Research
IT University of Copenhagen
Copenhagen
Denmark

ISSN 2197-9685   ISSN 2197-9693   (electronic)
Gaming Media and Social Effects
ISBN 978-981-13-3823-6   ISBN 978-981-10-1891-6   (eBook)
https://doi.org/10.1007/978-981-10-1891-6

Printed on acid-free paper

This Springer imprint is published by the registered company Springer Nature Singapore Pte Ltd.
part of Springer Nature
The registered company address is: 152 Beach Road, #21-01/04 Gateway East, Singapore 189721,
Singapore

Foreword

Open an important document on your computer. Select the entire text and then use the "cut" command to remove it all. Where is it?

*

Place as many important files as you can find into your computer's trashcan, recycle bin, or equivalent. Select the option to empty it. When the confirmation dialog appears, hover your mouse above the confirmation button and imagine clicking it. Feel the fear.

*

Hold your phone out the window or over the edge of a building or over a body of water. Don't let go.

*

What is the emoji for gamification? What is the emoji for the quantified self?

*

Use the stopwatch function on your phone to regulate your breathing. Breathe in for one second and out for one second. Don't stop.

*

Open your web browser in "private" mode. Don't navigate anywhere. Allow yourself to think private thoughts.

*

With your phone off, sit on a chair in a public space, and move your finger over the screen. Make sure you project emotions on your face. Smile wryly from time to time; mutter an expletive; roll your eyes. Make a decisive swipe, nod, stand up, and walk away.

*

Arrange the icons on your computer's desktop into a drawing of a smiling face. Smile at the smiling face. Feel that your computer likes you now.

*

Open the history of your desktop web browser. Scroll through the accumulated days and pages. Who are you?

*

Using a word processor, type random characters as fast as you can. Continue until you feel productive.

*

Ask Google a question. Keep asking different questions until you find one that yields no answers at all. Now answer the question.

*

Spend some quality time with your phone. Open your photo-taking app and turn on the front-facing camera. Sit on your sofa and sit your phone next to you. Watch television together. Make sure your phone doesn't fall asleep. Laugh together. Or, if you can, cry together.

*

It looks like you're writing a letter, but you're actually writing a poem.

Montréal, Canada   Pippin Barr

Contents

Editor and Contributors

About the Editor

Daniel Cermak-Sassenrath is Associate Professor at the IT University of Copenhagen (ITU), and member of the Center for Computer Games Research (game.itu.dk) and the Pervasive Interaction Technology Lab (PitLab, pitlab.itu.dk). Daniel writes, composes, codes, builds, performs and plays. He is interested in artistic, analytic, explorative, critical and subversive approaches to and practices of play. Discourses he is specifically interested in, are play and materiality, play and learning, and critical play. More information is available at www.dace.de.

Contributors

Viktor Bedö Tacit Dimension, HPI School for Design Thinking, Potsdam, Germany

Maggie Buxton AWHIWORLD, Whangarei, New Zealand

Pedro Luis Cembranos Madrid, Spain

Daniel Cermak-Sassenrath Center for Computer Games Research, IT University of Copenhagen, Copenhagen, Denmark

Chris Crawford Jacksonville, USA

William Drew London, UK

Mathias Fuchs Leuphana University, Gamification Lab, Lüneburg, Germany

Susanne Grabowski compArt: Center of Excellence Digital Art, University of Bremen, Bremen, Germany

Chad Habel Teaching Innovation Unit, University of South Australia, Adelaide, SA, Australia

Sidsel Hermansen Copenhagen, Denmark

Andrew Hope School of Social Sciences, University of Adelaide, Adelaide, SA, Australia

Margarete Jahrmann Game Design, Zürcher Hochschule der Künste, Zürich, Switzerland; Digitale Kunst, Universität für angewandte Kunst Wien, Vienna, Austria

Rilla Khaled Department of Design and Computation Arts, Concordia University, Montréal, QC, Canada

Sybille Lammes Leiden University, Leiden, Netherlands

Eva Mattes Brooklyn, NY, USA

Franco Mattes Brooklyn, NY, USA

Florian 'Floyd' Mueller Exertion Games Lab, RMIT University, Melbourne, VIC, Australia

Amani Naseem Copenhagen Game Collective, Copenhagen, Denmark

Michael Nitsche Georgia Institute of Technology/Digital Media, Atlanta, GA, USA

Tom Penney Playable Media Lab, RMIT University, Melbourne, VIC, Australia

Julian Priest Wellington, New Zealand

Samuel Van Ransbeeck CITAR | Centro de Investigação em Ciência e Tecnologia das Artes, Portuguese Catholic University, Porto, Portugal

Introduction

Our own period, which is transforming nature in so many and different ways, takes pleasure in understanding things so that we can interfere.

(Brecht 1964: 193, §46)

The diverse emerging practices of digital media appear to be essentially playful: users are involved and active; produce form and content; spread, exchange, and consume it; take risks; are conscious of their own goals and the possibilities of achieving them; are skilled; and know how to acquire more skills. They share a perspective of can-do, a curiosity regarding what happens next. Rokeby's (1998: 27) expectation that "the experience of culture can be something you *do* rather than something you are given" rings true.

This book starts with the proposition that digital media invite play and indeed *need* to be played by their everyday users. Early on, play turns out to be the paradigmatic form of using a computer. "This demonstration routine [of shooting pool] illustrates very dramatically the future role of the high-speed digital computer as the heart of a control system, or as a simulation device for complicated systems" (Bauer and Carr 1954[1]: 180).

Play is a grassroots movement. In play people do what they really want; they please nobody but themselves. It is pure and direct, unmediated, and true. Players follow their individual and collective aims, interests, and goals. Play with digital media questions, challenges, and redefines power relationships. "*Spacewar* was 'part of no one's grand scheme' and 'served no grand theory.' It was, Brand observed, 'heresy, uninvited and unwelcome,' yet also a 'flawless crystal ball of things to come' in computer use: 'interactive in real time,' graphic, encouraging user programming, 'a communication device,' promising 'richness and rigor of spontaneous creation and human interaction,' and 'delightful.' Spacewar announced 'computer power to the people' (Brand 1972)."

---

[1]I am grateful to Hans Dieter Hellige (artec, University of Bremen, FRG) for bringing the article to my attention.

Play implies control and ownership. Play is not given but taken. Often, a struggle between top-down and bottom-up control plays out. But the development we observe is not a unified movement, led by a single person or ideology. It happens not in an orderly manner, according to a grand plan or common purpose, but driven by opportunity and accident, reflecting multiple trends with often conflicting goals and approaches. New digital communities are forming, not replacing traditional social bonds as was assumed in the 1990s, but mixing with them. There is no easily identifiable separate class or subculture: groups are formed dynamically and disbanded again. People use the digital medium in many creative, provocative, and unintended ways; for instance, "communications and information technologies created for the military-security state were subverted into playful expressions of digital delight" (ibid.: 10, see Edwards 2013), and the eight-bit home computer era saw a surge in unlawful software pirating by teenagers (see Seeßlen and Rost 1984: 215). Play can be observed in social, economic, political, artistic, educational, and criminal contexts and endeavors. It is employed as a (counter)strategy, for tacit or open resistance, as a method and productive practice, and something people do for fun. Play is probably one of the most visible and powerful ways to appropriate the digital world.

In a nutshell, this book aims to define a particular contemporary attitude, a playful approach to media. It identifies some common ground and key principles in this novel terrain. Instead of looking at play and how it branches into different disciplines such as business and education, the phenomenon of play in digital media is approached unconstrained by disciplinary boundaries. The contributions in this book provide a glimpse of a playful technological revolution that is a joyful celebration of possibilities that the new media afford. This book is not a practical guide on how to hack a system or to pirate music, but provides critical insights into the unintended, artistic, fun, subversive, and sometimes dodgy applications of digital media.

Discussing playful interaction with digital media touches and moves between many ambivalent, conflicting, and even paradoxical positions. The following are some of these.

A convergence of play and (postmodern) life is observed as well as a widening hiatus. Play is an aesthetic experience in Kant's sense, that is, something seen from a disinterested standpoint, as itself, not *for* something. Play is complete, intrinsically motivated, free, purposeless, strong, and ignorant. It needs to take primary importance, at least temporarily. Play-external influences such as educational or political agendas endanger play and can reduce it to a mere pose or performance; the only genuine, pure, and full play is play that is played for its own sake. Play maintains and requires a strict (conceptual) border argues Huizinga. Players follow the internal logic of play, which often appears intransparent when observed from the outside. They do things that other people do not do or (and therefore do not) understand. Players choose to ignore or defy common sense, rationality, customs, rules, taste, policies, and regulations. "The playing human is half anarchist" (ibid.: 38, my transl.).

At the same time, as a practice and an event that is located and embedded in time and space and social situations, play is full of exchange, mirroring (and mocking) everyday life. Players regularly, actively, and authoritatively take control of (their) play and often are activated—creatively, productively, and critically—beyond play. Play effortlessly spills over into everyday life, in a very tangible sense and often in disruptive ways. Players act with much energy and abundant dedication and conviction. Play not only absorbs many resources but also creates new energies (Groos 1896: 15, 1899: 471 in Retter 2003: 11; Seeßlen and Rost 1984: 38). Play's fictions are quite indistinguishable from reality (Seeßlen and Rost 1984: 175), and who knows where play ends?

Users of digital media appear to believe themselves to be in control of the medium, and the computer is widely seen as the medium that finally hands control over the selection, production, and distribution of medial content back to its users. Alas, Wilson (1993) points out that "the inclusion of choice structures does not automatically indicate a new respect for the user's autonomy, intelligence, or call out significant psychic participation. ... The missing choices might be more important than the 'choices' offered." Far from being "automatically empowering and democratizing" (Dyer-Witheford and de Peuter 2009: 189), interactivity can be seen as "a cynical manipulation of the user, who is seduced by a semblance of choice" (Wilson 1993; see Huhtamo 2005 and Dunne and Raby 2001: 45), by giving him or her the illusion of control; instead of "overcom[ing] spectacle, ... it can be subordinated to, and even intensify, spectacular power" (Dyer-Witheford and de Peuter 2009: 189). Instead of being a path to (self-) liberation (e.g., Freire 2005 and Boal 2008), play becomes a placebo, decoy, or smokescreen.

On the other hand, play needs to be genuine if it is to be sustained. It rejects the idea of creating "microtopias" and includes "sensations of unease and discomfort rather than belonging" and "*sustains* a tension among viewers, participants, and context" as Bishop (2004: 70) observes is characteristic of art works by Thomas Hirschhorn and Santiago Sierra. A "simulation of lawlessness" (Kozlovsky 2007: 175) as facilitated by adventure playgrounds might not be enough. Players develop certain expectations, and they know and have experienced that they can change the course of events, conditions, and outcomes.

The computer strengthens and consolidates existing power relationships (Seeßlen and Rost 1984: 17, see Dyer-Witheford and de Peuter 2009: 5f.), but it also questions and challenges them (Seeßlen and Rost 1984: 14 and 211). In digital media, submission and subversion are closely related; "decentralisation has an unlikely twin ... centralized control" (Barney 2004: 65; see Dyer-Witheford and de Peuter 2009: 188–90). Interactive networked media are not only seen "as conducive to individual empowerment and an erosion of the power of centralized communication institutions and information distributors" and as a factor that puts "individuals – citizens, consumers, working people – ... increasingly in charge"; but also as using "increasingly sophisticated techniques of surveillance and control" that transport "the privatization and fragmentation of things like community, the public sphere, and various other sources of social solidarity" (Barney 2004: 67).

These are the same media that are used to protest as well as are used to discipline (see Hobsbawm 1989: 33). Play is seen as an effective way to direct and defuse people. Similar to an art installation that might be intended and described by the artist as being open-ended and inviting viewer emancipation (Bishop 2004: 68 on Rirkit Tiravanij's work), such a play situation might by its structure foreclose or prescribe participants' interaction with it, and "circumscribe … the outcome in advance" (Bishop 2004: 69); play becomes "thought control" (Pink Floyd 1979). The "success or failure" of the *Super Mario* game "depends on … the willingness of a player to identify, perhaps for hours, perhaps over the span of an entire lifetime, with a diminutive, running, jumping, red-capped plumber" (Dyer-Witheford and de Peuter 2009: 5).

Cultural artifacts including "new media representations" are made for specific uses and applications, from particular angles, and "are … always biased" (Mano-vich 2000: 40). Decisions have been made, and one thing has been done, and not another, and the design of playful experiences is no exception. "… game design is never an ideologically neutral process: games, as every other cultural product, reflect the designers' beliefs and value systems. And this is particularly visible in games that claim to 'simulate' actual non-deterministic situations" (Molleindustria 2015). From a postmodern perspective, there are no absolute positions, and there is no neutral.

But cultural artifacts are also blank with regard to meaning and routinely appropriated by their users in everyday uses. A designer's intention may only marginally influence the reception and effect of a work. "Meaning is produced within the systemic structure created by the artists only through the activity of the users involved" (Grau 2003: 306f.; see Winter 1995: 108 and Bishop 2004: 78), and potentially also without the structures provided. It is up to the people to make sense of things by implicit and explicit contextualization: "What readers do with a text depends on what is relevant for them in their social situation" (Winter 1995: 108, my transl.). For instance, a game such as *The Sims* can either be taken at face value and criticized for propagating a limited life vision and normative ideals, and rewarding mainstream behavior, such as "rampant consumerism" (Gonzalo Frasca qtd. in Soderman 2010), or to

> "deliver …its own political critique … as part of the gameplay. … The boredom, the sterility, the uselessness, and the futility of contemporary life appear precisely through those things that represent them best: a middle-class suburban house, an Ikea catalog of personal possessions, crappy food and even less appetizing music, the same dozen mindless tasks over and over—how can one craft a better critique of contemporary life? (Galloway 2006: 103f.)"

Readers dominate texts[2] (Winter 1995: 108); this is a basic property of artifacts (see Bishop 2004: 62 footnote 29 and Brey 1998). Humpty Dumpty is a typical user

---

[2] "Nicht der Text übt Macht über den Zuschauer aus, sondern das Publikum über den Text."

rather than a designer when he asserts (qtd. in Carroll 1871) that "when [he] use[s] a word ... it means just what [he] choose[s] it to mean – neither more nor less," and he declares himself "master" of this process.

Part I: Learning, Reflection and Identity

The first part of the book looks at play as a way to engage people, to offer new perspectives, insights, and experiences, and to change how they see the world. Notions of educational play, the relation between play and learning, and play as a reflective experience are discussed. Concepts that are referred to are the *experiment*, *exploration*, *risk*, and *surprise*.

Chapter "Questions Over Answers: Reflective Game Design" by Rilla Khaled advocates game design that promotes critical reflection on values, practices, and agency in relation to sociocultural, gender, religious, and economic forces as a means for change, action, and empowerment. She explores various perspectives on reflection and how it has (or has not) been designed for within games and inter-action design, before moving on to propose a new alternative design agenda from which to design, deconstruct, and understand play experiences.

The interplay of identity and people's engagement in online media is the focus of Tom Penney and Florian "Floyd" Mueller in their Chapter, "Playing the Subject." They observe how online media can essentialize identities through social sorting, creating positive feedback loops, and by commodifying niche communities; they present examples of online applications that are concerned with identity, and investigate how artists play with and subvert these constructs by playing many selves and producing caricatures.

When they created *PoliShot*, a political Dada game and interactive installation, Susanne Grabowski and Daniel Cermak-Sassenrath were confronted with the question of what is morally or ethically tolerable in digital games. The authors take their experiences as an occasion to enquire into and discuss the contradictions of the actual and the virtual: of concept and content. They see art as an interconnected and dynamic system consisting of the artist, the work, and its reception. The interactions in this system produce different and potentially conflicting meanings that vary over time and parties involved, and open challenging opportunities for play.

Chris Crawford describes how, with the passage of time, play behaviors evolved to become more complex and more closely attuned to specific behavioral needs, before play reached its apex of complexity in homo sapiens. He discusses the lessons that can be drawn from such an understanding for education and game design.

Dingbats Fucktory is a video installation by Pedro Luis Cembranos that shows thousands of *TrueType* dingbats which have been transformed and modified to subvert and comment on their original meaning. The dingbats form an ironic comment, carrying ambiguous and absurd messages. Their unintentional usage plays off and alludes to matters such as economics, politics, everyday life, and art. The 384 dingbats shown in this book are taken from the posters that accompany the installation.

Part II: System, Society, Empowerment

The book's second part is concerned with the relationship between play and society. Ideas about justice, creativity, rules, and community are discussed. Prominent issues are the military-entertainment complex, computer games and capitalism, intellectual property, censorship, and political activism, as well as the playful use of social and mobile networks.

Sybille Lammes examines the intricate relation between play and digital mapping in her Chapter "Destabilizing Playgrounds: Cartographical Interfaces, Mutability, Risk and Play." She looks at how playing with cartographical interfaces is a central and never neutral activity that invites users to change cartographic landscapes in playful and subversive ways, hence having the potential to change the very "nature" of maps and the spatial relations they invite people to produce.

In his Chapter "Crafting Through Playing," Michael Nitsche explores playing as a productive practice, drawing on concepts from craft research and game studies. He focuses on player-emergent production practices as they emerge from the players' own creativity. Connections from productive play to "critical making" are established by identifying the progression through process as being more relevant and interesting than a final product.

Playmakers in the Maldives is a collaborative work by Amani Naseem, William Drew, Viktor Bedö, and Sidsel Hermansen. The project involved ten international game designers collaborating with the Maldivian communities and individuals to create games and play events in the public spaces of the Maldivian capital island Malé. The chapter presents the games that were developed and describes some of the issues encountered during the project.

A Civilized Society is an uncommented collection of images by Eva and Franco Mattes. They asked anonymous crowdsourcing workers through an online marketplace to photograph themselves in poses of protest.

The Others by Eva and Franco Mattes is a slideshow of photos stolen from personal computers. Ten of the 10,000 images are shown in this book.

Part III: Mis-use, Struggle, Control

The third part of the book focuses on the misuse of digital media, the struggle for control, stepping outside accepted norms and behaviors, taking ownership, and getting away. There are numerous practices that can be seen either as creative use and skillful repurposing, or as actions that circumvent or break the law. Strategies and counter-strategies abound, both from above and below, for instance, copy-protection and ways to evade it. Further examples from the area of network and data security include hacking, code breaking, phreaking, and creating malware, viruses, or trojans. Digital media appear to invite these practices, and have the potential to turn everyday users easily into artists and criminals alike.

The repurposing of data is the topic of Samuel Van Ransbeeck's Chapter, "Sonification in an Artistic Context." He discusses the process of converting and subsuming initially useful data into a purely aesthetic experience, presents several projects, and aims to propose an aesthetic framework for sonification.

In their Chapter, "Little Big Learning: Subversive Play/GBL Rebooted," Chad Habel and Andrew Hope reject the notion of the passive learner and the use of static educational drill-based games, and propose the practices of game design and game creation as an approach to learning, drawing on anthropological notions of play. They explain how clear roles of teacher and student then begin to lose significance, and with them traditional power structures, social boundaries, and distinctions between the real and unreal.

Mathias Fuchs observes how rule structures and interfaces, inspired by computer games, are permeating modern society and are increasingly used by corporations to create and manage brand loyalty and to create value. His Chapter "Subversive Gamification," aims at stirring up commonsense notions of gamification as a marketing tool and discusses activist tactics, artistic concepts, and subcultural strategies in regard to a ludification of society.

Chapter "Constant beyond Gamification: Deep Play in Political Activism," by Margarete Jahrmann compares Geertz' ethnographic concept of Deep Play with current concepts of activist role play, social intervention, and public protest. The chapter finds its creative and intellectual leitmotif in "ludic" activist arts connected to contemporary forms of game arts and political role play.

Part IV: Place, Reality, Meaning

The final part of the book investigates the setting and place of play, and the relationship between play and ordinary life and reality. Topics include tangible interaction, mixed and augmented realities, tactile interfaces, haptics, motion-detection games, location-based play, the application of phenomenological ideas to interaction design, and the concept of embodiment.

In her Chapter "Tricksters, Games and Transformation," Maggie Buxton explores the relationship between trickster figures, emerging game formats and transformative learning theories. After describing today's relevance and role of tricksters, and how certain game formats can be seen as trickster tools, she posits a potential relationship between these tools and transformative learning.

The separation of play from the everyday world, the resulting conflicts, and the steps into and out of play are discussed by Daniel Cermak-Sassenrath in Chapter "Makin' Cake—Provocation, Self-Confrontation, and the Opacity of Play." An interactive installation demonstrates how play can act like a funhouse mirror into Wonderland, reflecting ordinary life but giving it its own twist, path, and, finally, meaning: free and independent of ordinary life.

Julian Priest's Free of Charge project draws together the playful aspects of the role-play in the security theatre, roles of authority and mischief, questions of belonging and ownership, and idea(l)s of convention and subversion. The participatory artwork is staged as a mock airport security check procedure that is modified to measure visitors' static electrical charge. Participants pass through the security checkpoint and are measured for charge before being electrically grounded and discharged. The chapter describes the work and its site and develops the rationale for the work. It is discussed in relation to the post-9/11 security apparatus and the concept of security theater, and this is contrasted with aspects of the work that deal with health and wellness around static electricity. Through these lenses response to authority and the internalization and subversion of roles are examined.

Chapter "Playing on the Edge," by Daniel Cermak-Sassenrath investigates how much subversion play can take. What are the boundaries of play and how far it can be pushed? Where is the edge and who negotiates it?

Acknowledgements

I would like to thank Charles Walker (Architecture & Future Environments, Auckland University of Technology, NZ) who had the initial idea for the book; and Chek Tien Tan (InfoComm Technology Cluster, Singapore Institute of Technology, SG) who reviewed several chapters early in the process.

References

Barney, D. (2004). *The network society*. Cambridge: Polity.
Bauer, W. F., & Carr, J. W., III. (1954). On the demonstration of high-speed digital computers. *Journal of the ACM, 1*(4), 177–182.
Bishop, C. (Autumn 2004). *Antagonism and relational aesthetics* (Vol. 110, pp. 51–79). MIT Press, www.jstor.org/stable/3397557 (08/04/2011)
Boal, A. (2008). *Theatre of the oppressed* (C. A. McBride, M.-O. Leal McBride, & E. Fryer, Trans.). London: Pluto Press.
Brand, S. (1972, December 7). Spacewar: Fanatic life and symbolic death among the computer bums. *Rolling Stone*. www.wheels.org/spacewar/stone/rolling_stone.html.
Brecht, B. (1964). *Brecht on theatre* (J. Willett, ed. and Trans.). London and New York: Methuen
Brey, P. (1998). The politics of computer systems and the ethics of design. In: J. van den Hoven (Ed.), *Computer ethics: Philosophical enquiry* (pp. 64–75). Rotterdam University Press, preprint version.

Carroll, L. (1871). *Through the looking-glass* (The Millennium Fulcrum Edition 1.7). https://www. gutenberg.org/files/12/12-h/12-h.htm. (April 30, 2016).

Dunne, A., & Raby, F. (2001). *Design noir: The secret life of electronic objects.* Basel/Boston/Berlin: Birkhäuser.

Dyer-Witheford, N., & de Peuter, G. (2009). *Games of empire. Global capitalism and video games.* Minneapolis: University of Minnesota Press.

Edwards, B. (2013, January 24). The never-before-told story of the World's first computer art (It's a sexy dame). *The Atlantic.* www.theatlantic.com/technology/archive/2013/01/the-never-before-told-story-of-the-worlds-first-computer-art-its-a-sexy-dame/267439 (April 30, 2016).

Freire, P. (2005). *Pedagogy of the oppressed. 30th anniversary edition* (M. B. Ramos, Trans.). New York: Continuum.

Galloway, A. R. (2006). *Gaming. Essays on algorithmic culture.* Minneapolis: University of Minnesota Press.

Grau, O. (2003). *Virtual art. From illusion to immersion.* (G. Custance, Trans.). Cambridge: MIT Press.

Groos, K. (1896). *Die Spiele der Tiere.* Jena.

Groos, K. (1899). *Die Spiele der Menschen.* Jena (Reprint Hildesheim 1973).

Hobsbawm, E. J. (1989). *The age of empire 1875–1914.* New York: Vintage.

Huhtamo, E. (2005). Slots of fun, slots of trouble: An archaeology of arcade gaming. In J. Raessens & J. Goldstein (Eds.), *Handbook of computer games studies* (pp. 3–21). Cambridge: MIT Press.

Kozlovsky, R. (2007). Adventure playgrounds and postwar reconstruction. In M. Gutman & N. de Coninck-Smith (Eds.), *Designing modern childhoods: History, space, and the material culture of children* (pp. 171–190). New Brunswick: Rutgers University Press.

Manovich, L. (2000). *The language of new media.* Available online www.manovich.net/LNM/ Manovich.pdf (August 9, 2002).

Molleindustria. (2015). *Oiligarchy Postmortem | Molleindustria.* Web site www.molleindustria. org/oiligarchy-postmortem (November 11, 2015).

Floyd, P. (1979). Another brick in the wall, part 2. LP. *The Wall.* (Written by Roger Waters).

Retter, H. (2003). *Einführung in die Pädagogik des Spiels.* Braunschweig: Institut für Allgemeine Pädagogik und Technische Bildung der Technischen Universität Braunschweig, Abteilung Historisch-Systematische Pädagogik, Erstdruck 1998, erweiterte Neuauflage.

Rokeby, D. (1998). The construction of experience: Interface as content. In C. Dodsworth, Jr. (Ed.), *Digital illusion. Entertaining the future with high technology* (pp. 27–47). Reading: Addison-Wesley.

Seeßlen, G., & Rost, C. (1984). *Pac-Man & Co. Die Welt der Computerspiele.* Reinbek bei Hamburg: Rowohlt.

Soderman, B. (2010). Every game the same dream? Politics, representation, and the interpretation of video games. *Dichtung Digital. A Journal of Art and Culture in Digital Media.* Nr. 40, 2010. www.dichtung-digital.org/2010/soderman/soderman.htm (April 30, 2016).

Wilson, S. (1993). *The aesthetics and practice of designing interactive computer events.* Available online http://www.online.sfsu.edu/~swilson/papers/interactive2.html (June 30, 2003). Published in a different form in ACM SIGGRAPH 93 Visual Proceedings Art Show Catalog, 1993.

Winter, R. (1995). *Der produktive Zuschauer. Medienaneignung als kultureller und äthetischer Prozeß.* Munich: Quintessenz, MMV Medizin.

Part I
Learning, Reflection and Identity

Questions Over Answers: Reflective Game Design

Rilla Khaled

Abstract Reflection is the mental process that occurs when we encounter situations that cannot be effectively dealt with using previous experiences and solutions. For decades, it has been acknowledged as an important process in learning, and in recent years it has become a central focus of branches of interaction design. Games are highly appropriate vehicles for triggering and supporting reflection, but several of the dominant tropes of conventional game design directly work against reflection. In serious games, the promise of *safe environments*, the drive to pose problems with *clear solutions* and a preference for *stealth learning* complicate how directly we can design for reflection. In mainstream entertainment games, qualities such as *immersion* and the design traditions of *designing for the everyplayer* and *quantifying motivation* again run counter to a reflective agenda. Drawing on the critical and reflective design literature and on case studies of experimental games on the peripheries of mainstream game design, I propose *reflective game design*, a new alternative design agenda from which to design, deconstruct and make sense of play experiences.

1 Introduction

Reflective thought has been defined by Dewey as, "active, persistent, and careful consideration of any belief or supposed form of knowledge in the light of the grounds that support it, and the further conclusions to which it tends" (Dewey 1933). It is the mental process that occurs when we encounter situations that cannot be effectively dealt with using previous experiences and solutions. Such situations lead us to revisit and reassess our previous beliefs intentionally and consciously in order to find solutions that make sense in the newly understood context. Reflection stems from a state of perplexity, surprise and doubt. A desire to make sense of this state is what motivates us to find novel solutions and framings (Dewey 1933; Solomon 1987).

For decades, reflection has been acknowledged as an important process in learning (Boud et al. 1985; Mezirow 1990; Solomon 1987). In recent years, it has become a

R. Khaled (✉)
Department of Design and Computation Arts, Concordia University, Montréal, QC, Canada
e-mail: rilla.khaled@concordia.ca

© Springer Nature Singapore Pte Ltd. 2018
D. Cermak-Sassenrath (ed.), *Playful Disruption of Digital Media*,
Gaming Media and Social Effects, https://doi.org/10.1007/978-981-10-1891-6_1

3

central focus of branches of interaction design, particularly in relation to the role of technology in our lives (Dunne 2006). More recently, this focus has moved beyond reflection *on* technology and has extended to reflection *through* technology (Sengers et al. 2005; Gaver et al. 2007).

Games are highly appropriate vehicles for triggering and supporting reflection. Games support the representation of situations, problems and belief systems. When playing a game, we expect perplexity and surprise. We expect to fail before we succeed. We do not necessarily expect that the problem-solving process will be easy, and we are prepared to look for evidence, perform analytical reasoning and look for patterns in exploring our way to new solutions. Games grant us agency to enact our proposed solutions and give us feedback to reflect on their consequences. Despite this initial foundation of support, several of the dominant tropes of conventional game design directly work against reflection.

In serious games, the promise of *safe environments*, the drive to pose problems with *clear solutions* and a preference for *stealth learning* complicate how directly we can design for reflection. In mainstream entertainment games, theoretically free to focus solely on entertainment, qualities such as *immersion* and the design traditions of *designing for the everyplayer* and *quantifying motivation* again run counter to a reflective agenda.

But neither of these game design movements were explicitly set up to support reflection. To gain insight into how reflection could potentially be incorporated into games, I look to the critical and reflective design literature seeking characteristic design qualities and strategies. I then examine two critically successful experimental games on the peripheries of mainstream game design that succeed in creating reflective experiences: Pippin Barr's *Art Game* Barr (2013) and Die Gute Fabrik's (2014) *Johann Sebastian Joust*. I analyse these games to establish how they create these experiences, incorporating insights into their designers on the place of reflection in their games and the role of reflection in game design more generally.

Drawing together these insights and highlighting design qualities of games that support reflection alongside qualities that hinder it, I propose *reflective game design*, a new alternative design agenda from which to design, deconstruct and make sense of play experiences.

2   Reflection, Learning and Games

Reflection involves rational analyses and scrutiny of the grounds of our beliefs (Dewey 1933). Mezirow points out that while critical thought is an implied feature of the reflective thought process, what we usually mean is *critical reflection*, an interrogative process in which we critically assess the validity of presuppositions on which our beliefs have been based or how problems are posed or defined in the first place (Mezirow 1990). Critical reflection is therefore less specifically focused on teaching us how to do, and more on how we make meaning, particularly concerning normative views, judgments, propositions, beliefs, opinions or feelings (Mezirow

1990). It is less focused on product and more focused on process. For the rest of this chapter, "reflection" will be taken to refer to critical reflection.

Dewey highlights that reflection is not necessarily an easy or comfortable process, as the analysis of existing beliefs requires a willingness to suspend judgement, and that suspension of judgement can be painful (Dewey 1933). Additionally, we cannot reflect unless the situation at hand calls to mind other relevant experiences or beliefs; no precedents mean there would be no beliefs to scrutinise in a new light.

Reflection has been acknowledged as an important learning process within the design community. This is most visible in the work of Schon via the concepts of *reflection-in-action* and *reflection-on-action*, in which designers reflect on the consequences of "moves" made during and after the design process to reconfigure their understanding of the design space (Schön 1983).

Reflection has also been profoundly influential on contemporary attitudes towards learning, featuring prominently in theories of learning such as constructivism (Piaget 1985) and experiential learning (Kolb 1984), as well as being embraced by contemporary thinkers on education (Mezirow 1990; Moon 1999; Solomon 1987). Within the learning technology community, we see the consequences of this influence, with significant effort being dedicated towards the development of tools that foster reflection in learners, e.g. (Chen et al. 2009; Cook et al. 2002; McNamara et al. 2006).

Examination of these tools reveals that they frequently make use of simulations, across diverse disciplines. For example, simulation-based reflective tools have been developed to support learning about construction project planning and control (Mawdesley et al. 2011), the regulation of calcium by the human body (Pilkington and Parker-Jones 1996) and teacher training (Yeh 2004). Simulations afford reflection in several ways. They provide the means to explicitly represent systems of beliefs, propositions and processes. These can in turn be instantiated by users, and manipulated and explored by them in ways enabling interrogations of validity. Notably, simulations usually afford the possibility of being run or experienced multiple times, with events and outcomes varying in accordance with user actions and inputs. This lends an external, tangible character to the qualities of reconsideration and reassessment necessary for the process of reflection (Boud et al. 1985; Dewey 1933; Mezirow 1990).

Games are highly related to simulations. Both are often used to model systems, situations and events. Both are essentially sequences of carefully designed experiences, co-created through the decisions of designers and our choices at run-time. As such, the potential for simulations to support reflection is similarly true of games. While games may be highly related to simulations, simulations are not necessarily synonymous with games: there are specific qualities that we associate with games that we do not with simulations. Unlike simulations, games are inextricably linked with the notion of designed challenge and often also with difficulty. We expect a game to present us with problems for which we may not have ready-to-hand or simple solutions. We expect them to confront us with situations requiring non-trivial effort on our part, requiring that we "step up our game". We analyse and leverage qualities of our in-game failures to move towards in-game successes. Furthermore, we appreciate hard-won game challenges and recognise that they are what trigger us to become stronger players. Non-trivial challenge, analysis and problem solving, key parts of the reflective process, are already present in how we generally understand games. To the extent that reflection is frequently characterised as the process through which we learn from experience (Boud et al. 1985; Schön 1983), games are reflection machines.

3 Reflection in Mainstream Game Design

Given this theoretical basis, we might therefore expect to commonly encounter reflection as a design quality and player experience in games. Yet reflection is in fact under-represented within both serious games and mainstream entertainment games. In the following, we will see that this relative absence stems from a fundamental conflict between certain conventions of mainstream game design and qualities of reflection.

3.1 Serious Games

As reflection has been highly influential on contemporary theories of learning and education, we might expect that reflection should be a core focus of *serious games*, games that focus on providing experiences alongside entertainment (Abt 1970; Winn 2008). Serious games have been developed to address a plethora of concerns and have leveraged diverse learning philosophies. *BirthPlay*, a training game that instructs players how to conduct breech deliveries (European Design Centre 2013) focuses on the formation of automated responses and reflexes as its intended learning objective. *Math Blaster* is a well-known example of a didactic pedagogy in serious game design, leveraging a drill-and-practice approach to learning arithmetic Davidson (1983). More in line with a reflective philosophy of learning, there are also serious games in which learning is positioned as an open-ended experience that requires player interpretation. *The Oregon Trail*, a now classic educational game, positions learners as pioneers travelling the Oregon Trail in the mid-1800s and learning about the conditions involved (Rawitsch et al. 1974).

At the same time, serious games have come to represent a subset of games smaller than that implied by its broad definition. From this, tighter subset emerges a set of design values that run counter to those that support reflection.

Safe Environments

An often-mentioned advantage of using serious games is that they provide "safe environments" for risk-free exploration of behaviours (Geurts et al. 2000; Hijmans et al. 2009; Raybourn 2000). The core rationale underlying this is that such environments enable players to experiment with behaviours that they may not otherwise be willing—or even able—to enact in real life. Additionally, behaviours enacted in-game are promised to have no consequences on related real-world situations. For example, a problematic delivery of a baby while playing *BirthPlay* has no real-world consequences. Safe environments are clearly an advantage games and simulations can provide for situations that, if enacted in the world, could result in danger or anxiety or would otherwise be too costly or difficult to reproduce. Safe environments may also be an important ethical consideration for games targeted at players at risk in various ways, for example, the very young, the frail or those at risk of exclusion. For example, *FearNot!*, a simulator that presents episodes of bullying, gets the player to propose coping strategies to the victim character, and has the player observe the consequences of the coping strategy while not needing to experience it from a first-hand perspective (Hall et al. 2009). But while the promise of safe environments increases the user-friendliness of serious games, at the same time it can render them innocuous.

If the simulated or fictional nature of a serious game is emphasised to the extent that players cannot and do not relate their in-game behaviours to the real world, it is unlikely that learning transfer will occur. For example, Hijmans et al. found that amongst students who played the management game *Lumière* over the course of 6 weeks, those who viewed it as "just a game" had significantly lower scores in terms of perceptions of instructiveness, challenge and a chance to test oneself, and higher scores in terms of boredom, than those who viewed the experience as being in "bitter earnest" (Hijmans et al. 2009). Those who went into their *Lumière* experiences viewing it as just a game benefited less.

Games allow players to take the driver's seat in shaping their own play experiences. Once the game is over, however, the memory of the play experience is what remains. Safe environments can pose problems for reflection in games indirectly because they privilege protecting players from having experiences that closely connect to real-world ones. If we always prioritise safe environments, we essentially muzzle the experiential capacities of games. Games that feel too safe can also feel irrelevant.

Solvability and Clear Solutions

While in games in general we expect to be faced with complex challenges that may be beyond our abilities, in serious games, challenges frequently have clear, "correct" solutions, often readily solvable by the average player. While this perhaps comes as no surprise for games on subject matter such as mathematics and physics—what is deemed a correct answer in a game about fractions should hold stable across players—it makes less sense for games on subject matter more philosophically inclined or subjective in nature—a game about empowering individuals on how to escape the conditions of homelessness should not have correct answers.

Ideological forces partly drive this bias: designing a serious game that is too hard for players to master or that cannot easily demonstrate player mastery may defeat its purpose of facilitating and showing learning, making it impossible to produce and measure learning effects. Technological forces similarly shape the bias: determining appropriate game challenges on the basis of predictive models of player knowledge remains a major undertaking for game AI. The models that can be built at this stage typically concern closed game challenges and problem-solving, for example, players' mastery of arithmetic fractions (Andersen 2012).

In simplifying game challenges down to the closed and measurable out of concern that players will otherwise not understand them, or indeed that the available technology cannot be used to model anything more complex, we miss out on the opportunity to use serious games to explore more open-ended, ambiguous and unclear problems. Arguably, presenting learners with straightforward, closed problem solving disempowers them, training them to expect that the domain content in question can only take specific, prescribed forms. It also under-exercises metacognitive learning skills, such as problem framing and synthesis, and does not acknowledge that much of the problem solving we conduct in the world takes into consideration complex interactions between interrelated systems. A game that seeks to empower individuals on how to escape the conditions of homelessness cannot responsibly claim to do so without addressing the murky web of factors involved relating to economics, cultural values, societal prejudices, institutional support (or lack thereof) and human agency, to name but a few.

Characteristic qualities of the reflection process include disruption, failure and surprise. It is these negative experiences that trigger the critical analysis necessary to reach new understandings that account for flaws in previous understandings. Games that privilege simple challenges do not create these moments of disruption.

Stealth Learning

Widely present in perspectives on serious games is the idea that learning should be disguised (Annetta 2008; Gee 2003). Prensky refers to this as *stealth learning*, a learning process in which learners become so immersed in playing educational games that they are not aware that they are learning while playing (Prensky 2001). In such games, non-entertainment objectives are essentially hidden under a veil of game engagement. Stealth learning rests on a series of assumptions about learning and fun: that learning is generally unengaging, but that games are fun, and therefore that games should be used to camouflage learning.

From a learning transfer perspective, there are serious concerns surrounding whether players are able to reapply knowledge and skills obtained via games if learning is intentionally masked. Experts in learning transfer propose that the application of contents and skills is most likely to take place when the associative strength between learned content and application context is higher than any other competing knowledge in memory (Fisch et al. 2005). In the context of stealth learning, if a game about spies conveys information about cryptography, but players do not associate this information with cryptography in a non-fictional context, then they are less likely to try to apply it in their lives. People need to closely associate learned content with a situation before they will apply it. Work on learning transfer in the psychology literature also shows that learning that takes place explicitly and consciously often results in the development of improved general problem-solving abilities and knowledge that can transfer to novel situations (Hayes et al. 1988; Mandler 2004). In the context of games specifically, Súilleabháin and Sime state that deep learning and transfer from games can only occur when there is a high degree of experiential fidelity presented in game (Súilleabháin et al. 2010).

Taken together, these results suggest that if players are unaware that they are learning, or what they are learning about, and if games do not provide a high fidelity representation of the context the learning content should be applied within, it is unlikely that players will embed these game experiences into their sense-making processes for addressing related problems in the world. In addition, players will likely be unable to extrapolate from their game experiences when faced with novel contextualisations of these problems. In a related vein, in the context of simulation gaming, Crookall and Hofstede et al. point out that post-game debriefs, and discussion sessions where the learning implications of games are explicitly addressed are an essential (and often overlooked) component of unpacking, contextualising and making sense of simulation gaming experiences. Games with no debriefed stage are generally less effective in terms of learning transfer than games that do have such a stage (Crookall 2010; Hofstede et al. 2010).

A lack of player awareness regarding learning may potentially make for a more entertaining experience. In terms of supporting reflection, however, it is likely to be more of a hindrance than a benefit. Stealth learning explicitly elides how play experiences relate to learning. Games that camouflage what is explicitly beneficial reduce the likelihood that players will leverage it in the world.

3.2 Entertainment Games

If values supporting reflection are not yet widely expressed within serious games for reasons related to pedagogical philosophies and user-friendliness, then perhaps we might expect that they should be more present in mainstream entertainment games in which learning is not the main objective. But for different reasons—this time related to dominant game design philosophies and current best practice in mainstream game design—again we will find that reflection ends up taking a back seat.

Immersion

One of the notable characteristics of games is their capacity for creating *immersion*. Indeed, immersion has generally been viewed as a desirable feature of game experiences, and one that has been wholeheartedly endorsed by the computer game industry (Salen 2004). While the term has been used in multiple ways, and in reference to a range of phenomena within the game studies literature (Murray 1997; Ermi 2005), Calleja suggests that at a basic level, agreed upon by most theorists, it connotes some sense of player involvement (Calleja 2011). This involvement, in turn, has most often referred either to a sense of absorption, for example, the experience of becoming engrossed in *Tetris* (Pajitnov and Tetris 1984) or to a sense of transportation, for example, the feeling of actually being in the world of a game like *BioShock* (Levine 2007). Of course, at any given time, both of these senses can play into a player's degree of involvement. In this work, immersion is referred to in the sense of transportation.

Immersion in games shares roots with a far older artistic tradition seeking to engender transportation in spectators, namely Aristotelian poetics (Frasca 2004). Theatre director Boal describes the experience of spectators in Aristotelian poetics as *catharsis*: in closely identifying and empathising with characters, we live vicariously through their experiences, and we allow what happens to them to serve as a cathartic experience for ourselves (Boal 1985). Worth noting is that rhetorician Burke proposes that identification is at the root of persuasion: "You persuade a man insofar as you can talk his language by speech, gesture, tonality, order, image, attitude, idea identifying your ways with his" (Burke 1950). That is, we are more able to be persuaded when we identify with the persuader. But Boal problematises this, arguing that such empathic identification absolves us from genuinely engaging with the real-world problems the fictional characters represent. A more insidious consequence, Boal claims, is that in empathising with characters, we implicitly accept the fictional constraints defining their situations. Supporting Boal's perspective with Burke's views on persuasion, if identification drives persuasion, and we feel empathy and close identification with characters, then we are persuaded to accept their perspectives and circumstances.

Boal claims that Aristotelian poetics lull us into accepting the *status quo* and are therefore not suitable for consciousness-raising efforts. As a more suitable vehicle for consciousness-raising, Boal presents the *Theater of the Oppressed*, theatrical spectacles representing real situations of oppression, to which spectators are invited to propose possible solutions through acting out victim roles. This form of theatre specifically eschews immersion: the intention is for many spectators to act out solutions to serve as, and fuel conversation. Frasca argues for adopting a similar approach in creating *Videogames of the Oppressed*, incorporating cycles of developing games expressing certain problems, perspectives and solutions, discussing game contents and play experiences, then redeveloping the games in the light of the ongoing conversation (Frasca 2004).

In debating and considering solutions to non-trivial, real problems, spectators in the case of the *Theater of the Oppressed* and players in the case of *Videogames of the Oppressed* require distance to think critically. Returning to reflection, while immersion may be desirable in the context of pure entertainment, it works against enabling us to consider our play experiences—in the moment at least—from an analytical perspective and critical distance. Immersion concerns being drawn into a fiction and experiencing a sense of convergence with it. Reflection, on the other hand, concerns introspection and active interrogation of beliefs, situations and persuasive claims, and demands critical distance. Games that maximise immersion do so at the cost of reflection.

Satisfying the Everyplayer

A visible trend that emerges when surveying computer games from the past three decades is that in recent years they have become easier (Abbott 2012). *Permadeath*, a non-recoverable death state common in games like *Rogue*, has now become the exception rather than the norm. *Save points* are now ubiquitous, radically changing the notion of death and failure in games to temporary states that need not signal game over or even much inconvenience. More forgiving gameplay forms part of a larger trend of designing to suit player expectations, in no small part motivated by the desire to appeal to wider audiences and thus to increase sales. Paradoxically, player expectations themselves are shaped by what is readily available on the market, like breeds like. But beyond financial explanations, designing towards player expectations and tastes looms large in the rhetoric of contemporary game design literature. In a widely used game design textbook, for example, Fullerton claims that "The role of the game designer is, first and foremost, to be an advocate for the player" (Fullerton 2008). Meeting player expectations is essentially prescribed as the "right" way to approach game design and practically a moral obligation on behalf of designers (Bateman and Boon 2006; Fullerton 2008; Schell 2008).

As Wilson and Sicart write in the context of *abusive game design*, in seeking to satisfy the desires of the everyplayer, "[t]he designer becomes the odd-one-out, pressured to efface their own presence in order to ensure that the game is optimally tailored to the player" (Wilson and Sicart 2010). Rough edges not conforming to mainstream game design norms or player tastes—potentially innovative or idiosyncratic of the designer—are sanded off lest they turn potential players away.

Designing to the needs of players is undeniably important and desirable in many situations, and can be a way to ensure that games are relevant, appropriate and meaningful. It can also be a way of introducing innovation and diversity into games (Khaled 2012). But designing to the needs of players need not mean that designer perspectives must be back grounded. Interaction design, for example, embraces cultural probes as a co-creative method that incorporates both designer and user perspectives, while retaining subjectivity of interpretation (Gaver et al. 1999). When looking for similar co-creation methods within mainstream game design practice, however, there is a paucity of support (Sotamaa et al. 2005; Sotamaa 2007).

Designing to satisfy the everyplayer can be indirectly problematic for reflection. It places designers in a service role, establishing a power hierarchy between designers and players in which designer perspectives rank lower than player perspectives. If designers are always concerned about meeting player expectations, then designing for experiences that deeply challenge the player, surprise them or trouble them may be discounted as possibilities. Designing for the everyplayer means that design becomes predictable and unchallenging.

Quantifying Motivation

A conventional way games communicate the desirability of certain actions is attributing points and achievements to them. In *Tetris*, clearing one line is worth 100 points, while clearing four lines at once is worth 800 points (Pajitnov and Tetris 1984). This approach can become confusing, however, in games that do not deal strictly with abstract themes, but rather with premises and challenges which we interpret using cultural knowledge. In these games, players may already have preconceptions about desirable actions to perform, influenced by social and cultural expectations governing appropriate conduct. In *Shadow of the Colossus*, for example, players must kill colossi to progress, yet killing them feels wrong (Ueda 2005). When these games encourage desirable actions via point collection, the place of the player's beliefs and values in guiding preferable conduct may be back grounded as the game's system communicates an in-built preference towards player decision-making that seeks to maximise progress.

Games seeking to promote ethical gameplay often pursue this kind of quantified approach; for example, in the *Fable* series, players collect "good" points for saving villagers, and "evil" points for breaking laws (Molyneux and Fable 2004). Sicart makes the argument that games that primarily load the communication of preferences surrounding players' ethical conduct into a game's mechanical system fail to be ethical or promote ethical thought (Sicart 2009). Instead, they enable a disengagement of action from its sociocultural meaning, as they invite players to approach game decision-making as a point maximisation exercise. In turn, this can blunt players' capacities for ethical thought within those game contexts. The social and cultural consequences of adopting certain actions become secondary, game actions become instrumentalised in service of winning, and motivation becomes quantified.

Encouraging action primarily through a game's mechanical system is problematic for reflection because it is not clear that it succeeds in making players genuinely reflect on game actions and challenges in the light of what they mean semantically and culturally, or in relation to their own life experiences. The actions that earned those points may well be meaningful and thought provoking, but, under the cover of points, what motivated those actions may become blurred. Games that quantify motivation distort the meanings of actions.

4 Reflection in Interaction Design

Design that foregrounds reflection appears to be under-represented by, and often at cross purposes with the prominent design conventions of mainstream game design, both for serious and entertainment games. Indeed, mainstream game design has not been evolving to explicitly support reflection. But such design does exist within the broad sphere of interaction design. Over the past decade, *critical design* and *reflective design* have both emerged as influential interaction design subfields. The core philosophies of these movements are examined next.

4.1 Critical Design

Critical design takes the position that design can be used as a critical medium to comment on the social, cultural, political and environmental impacts of technology (Dunne 2006). An example of critical design is Ernevi, Palm and Redström's *Erratic Radio*, a regular radio that also listens to electromagnetic fields emitted by active electronic appliances (Ernevi et al. 2007). As more electromagnetic fields are detected, the radio starts to detune. In Ernevi et al.'s words:

> Having listened to the radio for some time, you feel the need for some food. As you move into the kitchen, still trying to follow the radio program, it gets increasingly difficult to hear. As you pass the refrigerator and the freezer, the radio loses its channel completely, leaving you with just white noise to listen to. When moving the radio around in the kitchen, its sound reflects how strong the electrical magnetic field is at its current location. (Ernevi et al. 2007)

In using this radio, or radio-like object, we cannot help but become sensitised towards the presence of electromagnetic fields around us, and we are invited to reflect on our own power consumption patterns.

The *Erratic Radio* is not a "user-friendly" object: it intentionally subverts traditional notions we have about optimal radio listening experiences and challenges expectations we may have about consumer electronics being designed to assist ease of use. Subversion and provocation are idiosyncratic of critical design, which is less concerned with developing user-friendly designed objects and systems, and more with how design can function as a social commentary, stimulating discussion and debate among designers, industry and users about the quality of our electronically mediated lives (Dunne 2006). Dunne, an early proponent of critical design, challenges the place of user-friendliness in products altogether, arguing that it effectively enslaves us:

> Enslavement is not, strictly speaking, to machines, nor to the people who build and own them, but to the conceptual models, values, and systems of thought the machines embody. User-friendliness helps naturalize electronic objects and values they embody. (Dunne 2006)

In wanting to challenge underlying conceptual models, values and systems, design strategies to promote critical thought are called for, including provocation, user unfriendliness and disruption of expected technology behaviours. Drawing a parallel to poetry, Dunne points out that poetry is an "unfriendly" form of writing that focuses our attention on language, and argues that we should create disruptive technologies as poetry about technology.

Strongly present in Dunne's positioning of critical design is the role of aesthetics in technology and how under-explored it has been in comparison with technological progress and performance:

> The most difficult challenges for designers of electronic objects now lie not in technical and semiotic functionality, where optimal levels of performance are already attainable, but in the realms of metaphysics, poetry, and aesthetics, where little research has been carried out. (Dunne 2006)

Dunne's privileging of aesthetics makes sense in the light of the lineage of critical design, stemming from a tradition of art, which has traditionally been used as a vehicle for mounting commentary on our social, cultural, political and environmental assumptions and practices. But Dunne cautions that art often provokes to the point of alienation, inadvertently becoming difficult or impossible to relate to and thus irrelevant (Dunne 2006).

4.2 Reflective Design

Soon after critical design began to gain traction as a subfield, Sengers et al. proposed *reflective design* as an umbrella term for design practices shaped to support both designers and users in ongoing critical reflection about technology and its relationship to human life experiences (Dourish et al. 2004; Sengers et al. 2005). An example of reflective design is Gaver et al.'s sensor-based *Home Health Horoscope* system, which stems from a position of scepticism about whether sensor-based systems can build accurate representations of people's well-being. The system is partly comprised of a series of sensors placed in locations around the house, such as inside a kitchen cupboard, on a kitchen door and in a bay window love seat. Each of these locations corresponds to particular types of routine home activity; for example, the kitchen cupboard relates to cleaning. Sensor readings are then translated into a series of well-being metrics related to dimensions including "busy", "cheerful", "private" and "disordered". Each morning, the system identifies the two metrics that have changed the most, and outputs a "household horoscope" related to these metrics by concatenating two random sentences appropriate to those metrics on a ticker tape printer. As the assembled horoscope is intentionally ambiguous, system authority is undermined, and the responsibility of meaningful interpretation rests instead with the user (Gaver et al. 2007).

Like critical design, reflective design can be viewed as a form of intervention and as a vehicle for rethinking dominant metaphors and values. But whereas critical design tends to focus more on the experiences that arise from encountering objects and technologies designed to trigger critical thought, the purview of reflective design extends from designing for foregrounding reflection in our users to how designers can become more attuned to their own values and biases, as well as those embedded in design processes and technologies themselves (Sengers et al. 2005).

Reflective design draws on philosophies from various design practices including *participatory design*, particularly, recognition of design as a reflexive practice and the importance of acknowledging different agendas and perspectives within the design process (Muller 2003). It also takes inspiration from *value sensitive design*, foregrounding the role of values in design (while admittedly privileging the value of critical reflection) (Friedman and Kahn 2003). Reflective design also builds on critical design, although Sengers et al. note that if the provocations of critical design are not interpreted as hoped-for by their designers, then these designs can veer off into the ridiculous, lacking "footholds" for designers or users—ironically, the same criticism that Dunne makes of art (Dunne 2006).

Related to critical design, *ludic design* also has an influence on reflective design, proposing playfulness, curiosity and reflection as important design values (Gaver 2008). Centrally, ludic design involves a challenge to values of work and efficiency which until recently were taken for granted as desirable qualities in software and technology. Reflective design also draws on Schön's notion of *reflection-in-action* (Schön 1983). *Critical technical practice* also serves as an inspiration, particularly because it encourages the articulation of dominant metaphors that may be limiting or overly influencing our beliefs about design, and potentially hindering progress as well as understanding (Agre 1997).

For increasing the likelihood of users having reflective experiences, Sengers et al. propose three strategies. *Providing for interpretive flexibility* allows users to maintain control of and responsibility for the meaning-making process and building open-ended systems where reflection is necessitated. *Giving users licence to participate* provides digital scaffolding for bridging from the familiar to the unfamiliar and drawing on playfulness to encourage participation. *Providing dynamic feedback to users* gives users feedback on their interactions to provide a stimulus for reflection. For incorporating reflection into the design process, Sengers et al. propose three other strategies. *Inspiring rich feedback from users* makes evaluation and reflection an inherent part of the design, not merely a step added on at the end. *Building technology as a probe* uses technology as an experimental stimulus to understand users, the effects of technology, as well as reflecting on the practices of technology design and evaluation. *Inverting metaphors and crossing boundaries* turns traditional assumptions upside down and looks to practices that are left "un-designed for" as sources of inspiration.

5 Experimental Games

Within the experimental games scene, peripheral to mainstream games, it is possible to find games featuring many of the characteristics of critical and reflective design that also eschew many of the conventional game design tropes previously identified as problematic for reflection. It comes as little surprise that many games from this scene challenge and intentionally subvert typical game design conventions, and are aesthetically motivated and positioned in opposition to mainstream games. Much as Buxton discusses in his chapter "*Tricksters, Games and Transformation*" in connection with disorienting and disrupting playful technologies, these experimental games and designers often serve as "tricksters" in the world of digital games.

Arguably, however, experimental games have gone further than critical and reflective design in terms of triggering critical reflection because of the broader reach of games. Experimental games are often easily available and have a growing audience of players seeking them out. In addition, games—computer and otherwise—are deeply woven into the fabric of our social and cultural lives. Many people have literal and cultural access to games, while far fewer people have similar access to critical and reflective design. For example, it is trivially easy for many people to access a free Web-based game, while the same cannot be said for obtaining the *Erratic Radio*.

Here I present case studies of two experimental games that foreground in their design many of the same patterns found in critical and reflective design, while also introducing design forces specific to games and their distribution. These are *Art Game* by Pippin Barr, a game in which you play an artist making works for an exhibition (Barr 2013) and *Johann Sebastian Joust* by Die Gute Fabrik, a multiplayer motion controller-based party game (Die Gute Fabrik 2014). Both games have been highly critically successful, have in different ways managed to reinvent within their genres and importantly for our purposes create reflective experiences for their players. To gain a more balanced insight into the design intentions behind these works, I interviewed the designers of these games. Their perspectives are interwoven with the case study descriptions, presented below.

Art Game

In *Art Game*, the player assumes the role of one of the three possible artists: Cicero Sassoon, a painter, Alexandra Tetranov, a sculptor, or William Edge and Susan Needle, performance artists. *Art Game* invokes the "games as art" debate, presenting the proposition that the way in which we play games can be understood as art or in the very least appreciated aesthetically. Painting as Sassoon becomes a process of creation through playing the game *Snake*, sculpting as Tetranov takes the form of games of *Tetris*, and performance art as Edge and Needle is a playing of *Spacewar!* When the game begins, the player is greeted by the curator of MoMA, New York, who informs them that they will be participating in a new group show at the museum. The player's challenge, then, becomes to create artworks that are deemed worthy of exhibition by the curator. If the player succeeds in pleasing the curator, the player moves on to the exhibition itself, seeing their pieces on display in amongst other artworks and being experienced by an audience, as in Fig. 1.

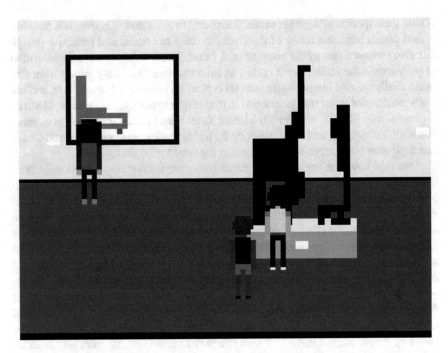

Fig. 1 An exhibition in *Art Game*

Art Game invites us to reflect on how we often fixate on using game mechanics instrumentally to pursue game objectives. In the game, it is never made clear what constitutes a desirable "*Snake* painting" or what kind of playing of *Spacewar!* will please the curator. The curator's impenetrable "art speak" complicates the notion that it is possible to decode, deconstruct and analyse her in order to give her what she wants. In fact, the curator's judgements are determined by random numbers. In conspicuously not providing a clear-cut path to success, and, in the words of Sengers et al. providing for interpretive flexibility, the game invites us to instead reflect on self-expression (see also, for example, Nitsche's work on chapter "Crafting Through Playing" of this volume). Barr states that part of his intention with *Art Game* and particularly in reappropriating classic game mechanics as art is to highlight that many games can be played expressively, beyond mere rule following:

> We're always creating these aesthetic experiences through play... A lot of games leave room for that kind of expression even if they don't validate it or express it. (P. Barr, pers. comm., June 05, 2013)

Like many of Barr's other works, *Art Game* is ultimately a game about games themselves. Barr triggers this reflection by subverting players' expectations and crossing the boundaries of certain rules of conventional game wisdom. One such rule questioned in several of Barr's games is that the designer should always provide the player with pleasantly entertaining experiences. On the topic of boredom and irritation as an intended design quality of *The Artist is Present*, a prequel to *Art Game*, Barr says:

> I tried to make the game have all these irritating realities embedded in it, like the museum having opening hours and a queue... the normal idea of usability and playability would elide these kinds of realities because they're boring, and the player shouldn't have to deal with boring things. Having those realities in there made the game more interesting and more intense than if they hadn't been there. I think it successfully made the point that it's not necessarily a safe assumption that you should make everything straightforward and easy and that you should elide boring things, because boring things can be very interesting. (P. Barr, pers. comm., June 05, 2013)

Designing for boredom in games in many ways mirrors the theme of "user unfriendliness" advocated for in critical design. But a potential critique of user unfriendliness is that it is merely a knee-jerk reaction to mainstream design, only fleetingly interesting because it runs counter to convention. Barr says the following on how *not* providing players with conventionally satisfying experiences can lead to them having deeper experience-triggered insights about games:

> It's good if there are games that are difficult or grate against the player's expectations in such a way that part of what they think when they're playing is hopefully not just "this game is annoying" but also "why is this game like this? This person probably didn't just make this game to be annoying" and so that can serve as a kind of trigger, some kind of irritation, that can then jump from just being annoying to... not necessarily a realisation but at least a few thoughts about "why does this game feel this way?" (P. Barr, pers. comm., June 05, 2013)

Such a reading of *Art Game* is not possible unless players already have some degree of game literacy. You need to know games to "get" *Art Game*. Barr also points out, though, that his intention is not just to be in conversation with an elite core of gamers with hard-won game expertise, but to reach a broader audience who only needs to possess a passing knowledge of games. He wants to give players licence to participate:

> Everyone knows about play and everyone knows about games in the general sense of something with rules where you're trying to do something. So everyone can understand jokes about those sorts of experiences, whether or not they are literate in digital games specifically. I think everyone has a picture in their minds of games and their fairness and these sorts of things. I'm making games more about that level of games rather than about video games and maybe that's why it's more accessible. (P. Barr, pers. comm., June 05, 2013)

On the subject of how mainstream hardcore gamers responded to *Art Game*, an audience more likely to be wedded to conventional game design, Barr says:

> They've generally been extremely positive and they've generally seen the humour in what I was doing and got the joke that was being played out. That's in no small part because I don't charge anything for my games and maybe if some of these people had paid for my games they would be irate about it. (P. Barr, pers. comm., June 05, 2013)

Barr thus highlights a curious tension: we can get away with surprising or irritating players and also have them view these experiences positively if these games are easily available and come at no (other) cost.

In summary, *Art Game* features several of the design qualities and strategies of critical and reflective design. The game deals with unusual subject matter for games—namely making art. It challenges conventional game wisdom regarding the instrumental use of mechanics, and how game feedback should facilitate progress, shifting interpretation back onto the player. Barr explicitly seeks to give players licence to participate in making his games accessible to those with only limited game experience. Finally, he uses his games as experimental stimuli, provoking questions in the minds of players about design conventions, as well as using them to probe what his players are willing to do.

Johann Sebastian Joust

In *Johann Sebastian Joust* (aka *Joust*), two or more players each hold a motion-sensitive controller and compete to be the last player "standing". The software and hardware of *Joust* enforce no rules other than this: if the motion of a player's controller is detected to be above the threshold determined by the tempo of the music playing, the game is over for that player. Players can therefore be eliminated for any number of reasons, including dancing too fast, dodging, being kicked by another player or even accidentally tripping over. To emphasise a sense of shared co-located play with a focus on other players, the game features no screen-based visuals at all, with only music of a varying tempo to guide player action. When the music is slow, the controllers are extraordinarily sensitive to motion. When the music is faster, the sensitivity is less so. Players act and strategise accordingly.

Joust subverts existing conventions regarding the design of games that make use of motion controllers by eliminating the use of screen and its corresponding virtual world, focusing visual attention on the physical players themselves. It creates dynamic feedback loops to players both through the hardware it uses and the reactions of other players and spectators. The game also places a strong emphasis on certain aspects of play in computer games that are often overlooked: extremes of physicality and the role of music. Innovation through the subversion of existing conventions is central to why *Joust* stands out, and became a critical success. By playing with game design traditions, *Joust* invites reflection on what we expect of typical motion controller games and how we expect players to conduct themselves. Doug Wilson of Die Gute Fabrik proposes the following on how the game succeeds in creating this response in players:

> I think there's more of a moment of humour where they kind of understood, maybe not thinking about it super clearly but "that's really neat" or "that's kind of innovative, that's unexpected" and their excitement is higher, now they're a little bit more into the game, now they wanna tell people about it, or play it more vigourously you can't feel some of that humour if you didn't know there were conventions, there would be nothing to subvert. The subversion implies that you recognise that there were traditions that it was different from. (D. Wilson, pers. comm., June 08, 2013)

Mirroring the theme of interpretive flexibility from reflective design, *Joust* has a very open system in terms of hardware and software rule enforcement, placing the onus on the players to come up with their own rules and interpretations, challenging assumptions around the role of game technology in upholding rules. Instead, as Wilson notes, players must be creative not just with their play, but with the rules of play themselves:

> All the time people come up with weird improvisations or physical improvisations that are kind of clever, that's the nice thing about a game like that with a somewhat open system, like the whole point of the game is trying to get people to make up their own house rules and bring something of their own to the game. I would say that that's in the very DNA of the design approach. (D. Wilson, pers. comm., June 08, 2013)

As players quickly discover, *Joust* does not reward "user-friendliness". It is entirely possible to play this game in physically extreme ways, and this can rapidly become an arms race of increasing forcefulness before house rules emerge with a stabilising effect (if such rules emerge at all). *Joust* is provocative, pushing the boundaries of players' physical and social comfort—which for some players results in them choosing *not* to play. As Wilson says:

> Certainly the game isn't for everyone. Some people don't play it as much, or play once and don't play it again, or don't act theatrically, some people really enjoy it... The game isn't for everyone and that's okay as well. (D. Wilson, pers. comm., June 08, 2013)

In walking this fine line of provocation, *Joust* risks alienating audiences not enamoured of highly physical or theatrical play. In the context of experimental games, this is not necessarily problematic. In the context of critical and reflective design, though, it anecdotally serves Dunne's point that some provocation is good for triggering critical reflection, but too much can alienate the audience (see also Grabowski and Cermak-Sassenrath's discussion of the game PoliShot in their chapter "*The Potential of the Contradictory in Digital Media—The Example of the Political Art Game PoliShot*").

While *Joust* might not be to everyone's taste, in other ways the game is quite inclusive, not requiring players to be hard core gamers, nor to have a particularly well-developed game literacy to understand why the game is subversive or appreciate how it is innovative. With regard to *Joust* and *B.U.T.T.O.N.*, another experimental game co-designed by Wilson that also riffs on physicality in games and the role of technology in upholding rules, Wilson says:

> [B]oth games work on the subversion of the technology, even beyond the gaming literate people, like you don't usually play video games without a screen, or like you don't usually play video games that become like a sport where you push each other or physically interact with each other. (D. Wilson, pers. comm., June 08, 2013)

Joust stands out because it uses motion controller technology in ways that oppose what we would expect. But central to its success is the fact that it makes use of a widely accessible motion controller technological platform. Wilson reminds us that distribution is a central, pragmatic and often downplayed factor in reaching the public via games:

> The thing that is super key is that it's done with consumer hardware... you can imagine that if I had built *Joust* in a research lab somehow slapped together with hardware from the research lab, I don't think the game would have played out the same way, it wouldn't have been so distributable, partly because I do think there's a "woah, you're totally using this PlayStation controller for an unintended purpose that's really neat". There's that kind of spark to people's imaginations. (D. Wilson, pers. comm., June 08, 2013)

Wilson maintains that designs that promote reflection must resonate with players, and above all be memorable:

> If your mind and heart isn't caught up in it, is it going to stick with you? Design for reflection is impossible without that "sticking with you"-ness. You can kind of see the shape of that in *Joust* with memorable play ... and it's the thing I care about the most ultimately—and I think it's probably what a lot of people care about—I think, "what games live with me afterwards?" (D. Wilson, pers. comm., June 08, 2013)

Joust, too, carries many of the trademarks of critical and reflective design. It subverts motion controller game design conventions by doing away entirely with the screen and enabling physically extreme play. It invites dynamic feedback loops, requiring players to look at—and to—each other, creating theatrical situations and enabling even non-player spectators to contribute towards house rules. The openness of the rule system enables interpretive flexibility, allowing the game to take on radically different flavours depending on the players and surrounding social context. In repurposing the constraints of familiar consumer hardware, and creating a play situation that is both familiar and unfamiliar, *Joust* also gives players licence to participate. In reflecting on *Joust*, Wilson proposes another core characteristic of reflective design, in keeping with Dewey's original notion of reflection: it must be memorable.

6   Reflective Game Design

Reflection involves revisiting and reassessing previous beliefs intentionally, consciously and carefully (Dewey 1933; Mezirow 1990). It relies on us being introspective about our beliefs and knowledge structures, and is associated with deep forms of learning (Moon 1999). Games theoretically afford reflection in several ways. They are experiential systems, whether these systems are enforced by hardware, software or social and cultural rules. Games can be used to explore and represent the same kinds of belief structures we draw on during reflection, giving us ways to reconsider, revise and reflect on existing patterns and assumptions, and potentially the agency to propose alternative solutions and framings. We expect and welcome failure and challenge in games, these being important components of the reflection process. But when looking for reflection within mainstream game design approaches, reflective design turns out to be rather under-represented.

Within serious games, which on the surface seem a natural fit for reflective agendas, expectations regarding the provision of safe environments mean that serious game experiences often play it *too* safe and end up bearing little relevance to related real-world situations and having little long-term effect on us. The pressure to provide players with solvable problems means that serious games rarely tackle genuinely complicated and ambiguous situations, and under-exercise our metacognitive and synthesis skills. Finally, the pervasive stealth learning perspective means that many serious games are designed specifically to obfuscate how play experiences connect to the world: players are not invited to reflect, as this would be too recognisable as learning.

Within mainstream entertainment games, immersion has been embraced to the detriment of reflection, serving almost as its antithesis. The notion of the every-player looms large within conventional game design practice, inadvertently curtailing game designers' freedom to explore less conventional avenues. In addition, with so many game systems designed around rewarding preferable actions with points and achievements, players' motivations to act become quantified, pushing them to act in mechanically beneficial ways, rather than to reflect on the social and cultural meanings of their actions.

Thus, in spite of games seeming like they should be a good fit for enabling reflective experiences, many of the characteristic design tropes of mainstream games work at cross purposes to reflection.

In contrast, within interaction design reflection has been actively pursued as a core design goal. In critical design, user unfriendliness is called for over user-friendliness in an attempt to grab the user's attention and trigger critical reflection on norms governing our use of technology. In reflective design, providing for ambiguity of interpretation and multiplicity of use is advocated, complicating the notion that a given product should have an easily determinable function. Inverting traditional assumptions and using as inspiration "un-designed for" practices is encouraged, allowing for questioning of the norm of seamlessness in our assumptions and expectations of how technology should work and why. Within both these design agendas, common design strategies and approaches involve playing off widely accepted design traditions, essentially attempting to create moments of surprise, a hallmark characteristic of reflective experience.

Within experimental game design, where designers have generally focused less on mass market appeal and more on design innovation through the subversion of mainstream game design conventions, we find some games that privilege reflection, mirroring the reflection strategies advocated for in critical and reflective design. In Barr's *Art Game*, reflection concerns a shift to viewing playing games as art creation, and how we can always use game mechanics expressively. Familiar mechanics are repurposed for making art in the game, and we are invited to be both playful and artful in our use of these mechanics. In Die Gute Fabrik's *Johann Sebastian Joust*, reflection concerns how we typically think about motion sensor games and technology, play as performance and spectatorship, and the place of house rules in computer games, as rules are revealed to be highly under-designed in the game's mechanical system. Amongst other reasons, *Art Game* and *Johann Sebastian Joust* have been highly successful *because* they subvert our assumptions surrounding more conventional game design.

In short, as critical and reflective design suggests, and as some experimental games show, games *can* successfully create and foster experiences for players that trigger critical reflection. This reflection may concern sensitising players towards underlying assumptions and values inherent in familiar systems, and provoking them into deeply exploring, questioning and co-creating responses to problems in the light of their own experiences and beliefs. Designing for surprise, player unfriendliness, ambiguity and multiple interpretation can push players towards reflecting on their play experiences, as can building gameplay around broken and recycled mechanics and open systems. Designing games for reflection involves rethinking the boundaries of what constitutes a game.

Games that promote reflection have, to this point, often supported reflection on conventions surrounding mainstream game design. But the reflective scope of these games need not solely concern reflection on games. Their reach can extend beyond—to social, cultural, religious and political values, practices, forces and systems, for supporting empowerment and serving as an agent of change. Such games would invite their players to harness their individual reflective and decision-making capacities in ways that acknowledge, problematise and potentially conflict with existing cultural and social norms. In doing so, they would provoke and challenge players.

I name the practice of designing such games *reflective game design*.

I close by proposing an agenda which lays out design qualities of reflective game design de-emphasising certain conventionally accepted design qualities in favour of more reflective ones.

Questions Over Answers

Reflection is not about answers or reaching "correct" solutions. It concerns deep consideration of problem spaces and is premised on questioning and revisiting our existing assumptions. Games that promote reflection will be, therefore, less about providing players with clear-cut, singular solutions, and more about creating opportunities for players to explore multiple possibilities and re-imagining problem framings. Asking meaningful questions is more important than providing clear answers.

Games that prompt questions invite players to be both introspective and proactive. They demand that players become more self-aware and critical about their relation to games, and more thoughtful about their capacity for critical reflection, action and individual agency. These games empower players, asking them to serve as critical commentators on their own experiences and to take ownership over them. They require players to engage at the meta-level, rather than merely reacting at the immediate level.

Clarity Over Stealth

Stealth learning may be enjoyable and a necessary enticement for players opposed to the concept of games with utility beyond entertainment. But games designed to promote reflection are not aimed at those audiences—or indeed, those designers who communicate preferences for stealth learning through their games. They are for those who *want* to know how to contextualise their game experiences back to the world, and who welcome the possibility of game experiences with complex meanings that endure beyond play. Upholding reflection means privileging clarity over stealth.

In focusing on clarity, games designed to trigger reflection promote *conscious* learning in contrast to accidental learning. Players *know* how and why they learned something from a game, because the game will not have been designed to obfuscate this. Players are supported in focusing on real-world connections, in order to maximise the chance that game-derived knowledge will not be segregated with "just a game" experiences but integrated with knowledge we use in daily life.

Disruption Over Comfort

Reflection is triggered when we are *not* strictly comfortable, when our assumptions are thrown into question and when we are confronted by situations that challenge our *status quo*. Games promoting reflection seek to create moments that lead to disruption and thus embrace designing for surprise, awkwardness and uncertainty. Disruption is more likely to lead to reflection than comfort.

Games that are designed to disrupt can create opportunities for players to be thoughtful, creative and innovative. They invite players to take on new ways of problem interpretation and solving, providing contexts that necessitate solutions beyond the obvious. Perhaps most importantly, disruptive game experiences stand out. If the intention is for players to reflect on how game experiences connect to the world, and to continue reflecting on this after play, these experiences must be memorable.

Reflection Over Immersion

Immersion is a highly desirable quality if design objectives mainly concern escapism. But reflection is precisely *not* about escapism: it concerns revisiting our previous beliefs intentionally and with a high degree of self-awareness. In the context of games, it requires acknowledging and incorporating the "fourth wall", even if this conflicts with the experience of "being there". Supporting reflection in games calls for privileging reflection over immersion.

Deeply reflecting on a game experience requires engagement with levels of game understanding and complexities of insight approaching that of the game's designer. Games that promote reflection demand active interpretation from their players as opposed to the passive consumption of experiences; indeed, post-play, players are in a position to contribute towards the game's redesign. If immersion is a quality we associate with playing—and losing ourselves in—games, then reflection is closer to *finding* ourselves in games and designing our own experience.

References

Abbott, M. (2012). Video games are easier than ever, yet harder to manage. https://kotaku.com/5887020/video-games-are-easier-than-ever-yet-harder-to-manage.

Abt, C. C. (1970). *Serious games*. Viking Press.

Agre, P. (1997). *Computation and human experience. Learning in doing*. Cambridge University Press.

Andersen, E. (2012). Optimizing adaptivity in educational games. In *Proceedings of the 7th International Conference on the Foundations of Digital Games* (pp. 279–281). ACM Press. https://doi.org/10.1145/2282338.2282398.

Annetta, L. A. (2008). Video games in education: Why they should be used and how they are being used. *Theory Into Practice: New Media and Education in the 21st Century, 47*(3), 229–239.

Barr, P. (2013). Art Game. http://www.pippinbarr.com/games/artgame/ArtGame.html.

Bateman, C., & Boon, R. (2006). *21st Century game design. Charles river media game development series*. Charles River Media.

Boal, A. (1985). *Theatre of the oppressed*. Theatre Communications Group.

Boud, D., Keogh, R., & Walker, D. (1985). *Reflection: Turning experience into learning*. Kogan Page.

Burke, K. (1950). *A Rhetoric of motives*. Prentice-Hall.

Calleja, G. (2011). In-game: From immersion to incorporation. MIT Press.

Chen, N. S., Wei, C. W., Wu, K. T., & Uden, L. (2009). Effects of high level prompts and peer assessment on online learners' reflection levels. *Computer Education, 52*(2), 283–291. https://doi.org/10.1016/j.compedu.2008.08.007.

Cook, J., & Oliver, M. (2002). Designing a toolkit to support dialogue in learning. *Computer Education, 38*(1–3), 151–164. https://doi.org/10.1016/S0360-1315(01)00076-8.

Crookall, D. (2010). Serious games, debriefing, and simulation/gaming as a discipline. *Simulation Gaming, 41*(6), 898–920. https://doi.org/10.1177/1046878110390784.

Davidson (1983). Math Blaster!.

Dewey, J. (1933). *How we think: A restatement of the relation of reflective thinking to the educative process*. Heath and company: D.C.

Die Gute Fabrik (2014). Johann Sebastian Joust. http://www.jsjoust.com/.

Dourish, P., Finlay, J., Sengers, P., & Wright, P. (2004). Reflective HCI: Towards a critical technical practice. In: CHI '04 Extended Abstracts on Human Factors in Computing Systems, CHI EA '04, pp. 1727–1728. ACM, New York, NY, USA. https://doi.org/10.1145/985921.986203.

Dunne, A. (2006). *Hertzian tales: Electronic products, aesthetic experience, and critical design*. The MIT Press.

Ermi, L., & Mäyrä, F. (2005). *Fundamental components of the gameplay experience: Analysing immersion*. In *DIGRA*. DIGRA.

Ernevi, A., Palm, S., & Redström, J. (2007). Erratic appliances and energy awareness. *Knowledge, Technology, and Policy, 20*(1), 71–78.

European design centre: Birth play (2011–2013).

Fisch, S., Kirkorian, H., & Anderson, D. (2005). Transfer of learning in informal education: The case of television. In J. P. Mestre (ed.) *Transfer of learning from a modern multidisciplinary perspective*, pp. 371–393. Information Age Publishing.

Frasca, G. (2004). Videogames of the oppressed: Critical thinking, education, tolerance, and other trivial issues. (pp. 85–94). MIT Press, Boston. http://www.electronicbookreview.com/thread/firstperson/Boalian.

Friedman, B., Kahn Jr., P. H. (2003). The human-computer interaction handbook. chap. Human values, ethics, and design, (pp. 1177–1201). L. Erlbaum Associates Inc., Hillsdale, NJ, USA.

Fullerton, T. (2008). Game design workshop, Second Edition: A playcentric approach to creating innovative games (Gama Network Series), 2 edn. Morgan Kaufmann.

Gaver, B., Dunne, T., & Pacenti, E. (1999). Design: Cultural probes. *Interactions, 6*, 21–29. https://doi.org/10.1145/291224.291235.

Gaver, W. (2008). Designing for homo ludens, still. In Binder, T., Lowgren, J., Malmborg, L. (eds.) (Re)Searching the Digital Bauhaus. Springer.

Gaver, W., Sengers, P., Kerridge, T., Kaye, J., & Bowers, J. (2007). Enhancing ubiquitous computing with user interpretation: Field testing the home health horoscope. In *Proceedings of the SIGCHI Conference on Human Factors in Computing Systems*, CHI '07, (pp. 537–546). ACM, New York, NY, USA. https://doi.org/10.1145/1240624.1240711.

Gee, J. P. (2003). *What video games have to teach us about learning and literacy*. Palgrave Macmillan.

Geurts, J., de Caluwé, L., & Stoppelenburg, A. (2000). *Changing Organisations with gaming/simulations*. Elsevier Bedrijfsinformatie.

Hall, L., Jones, S., Paiva, A., Aylett, R. (2009). Fearnot!: providing children with strategies to cope with bullying. In *Proceedings of the 8th International Conference on Interaction Design and Children*, IDC '09, (pp. 276–277). ACM, New York, NY, USA. https://doi.org/10.1145/1551788.1551854.

Hayes, N. A., & Broadbent, D. E. (1988). Two modes of learning for interactive tasks. *Cognition, 28*(3), 249–276.

Hijmans, E., Peters, V., van de Westelaken, M., Heldens, J., van Gils, A. (2009). Encounters of a safe environment in simulation games. In Bagdonas, E., Patasiene, I. (eds.) Games: Virtual Worlds and Reality. Selected papers of ISAGA 2009.

Hofstede, G. J., De Caluwé, L., & Peters, V. (2010). Why simulation games work-in search of the active substance: A synthesis. *Simulation Gaming, 41*(6), 824–843. https://doi.org/10.1177/1046878110375596.

Khaled, R. (2012). Muse-based game design. In Conference on Designing Interactive Systems, (pp. 721–730). ACM.

Kolb, D. (1984). Experiential learning: Experience as the source of learning and development. Prentice-Hall.

Levine, K. (2007). BioShock. 2K Games.

Mandler, J. (2004). *The foundations of mind: Origins of conceptual thought*. USA: Oxford Series in Cognitive Development. Oxford University Press.

Mawdesley, M., Long, G., Al-jibouri, S., & Scott, D. (2011). The enhancement of simulation based learning exercises through formalised reflection, focus groups and group presentation. *Computers & Education, 56*(1), 44–52. https://doi.org/10.1016/j.compedu.2010.05.005.

McNamara, D. S., O'Reilly, T. P., Best, R. M., & Ozuru, Y. (2006). Improving adolescent students' reading comprehension with istart. *Journal of Educational Computing Research, 34*(2), 147–171.

Mezirow, J. (1990). *Fostering critical reflection in adulthood: a guide to transformative and emancipatory learning*. Jossey-Bass Social and Behavioral Science Series: Jossey-Bass Publishers.

Molyneux, P. (2004). Fable. Microsoft Game Studios.

Moon, J. (1999). *Reflection in learning and professional development: Theory and practice*. Taylor & Francis.

Muller, M. J. (2003). The human-computer interaction handbook. chap. Participatory design: the third space in HCI, (pp. 1051–1068). L. Erlbaum Associates Inc., Hillsdale, NJ, USA.

Murray, J. H. (1997). *Hamlet on the Holodeck: The future of narrative in cyberspace*. New York, NY, USA: The Free Press.

Pajitnov, A. (1984). Tetris. Infogrames.

Piaget, J. (1985). *The equilibration of cognitive structures*. Chicago, IL: University of Chicago Press.

Pilkington, R., & Parker-Jones, C. (1996). Interacting with computer-based simulation: The role of dialogue. *Computers & Education*, *27*(1), 1–14. https://doi.org/10.1016/0360-1315(96)00013-9.

Prensky, M. (2001). *Digital game-based learning*. McGraw-Hill.

Rawitsch, D., Heinemann, B., Dillenberger, P. (1974). *The Oregon trail*. Minnesota Educational Computing Consortium.

Raybourn, E. M. (2000). Designing an emergent culture of negotiation in collaborative virtual communities: The case of the domecitymoo. *SIGGROUP Bulletin*, *21*(1), 28–29. https://doi.org/10.1145/377272.377293.

Salen, K., Zimmerman, E. (2004). *Rules of Play: Game design fundamentals*. MIT Press.

Schell, J. (2008). The art of game design: A book of lenses, 1 edn. Morgan Kaufmann.

Schön, D. A. (1983). *The reflective practitioner: How professionals think in action*. Basic Books.

Sengers, P., Boehner, K., David, S., & Kaye, J. J. (2005). Reflective design. In: Proceedings of the 4th decennial conference on Critical computing: between sense and sensibility, CC '05, (pp. 49–58). ACM, New York, NY, USA. https://doi.org/10.1145/1094562.1094569.

Sicart, M. (2009). The banality of simulated evil: Designing ethical gameplay. *Ethics and Information Technology*, *11*(3), 191–202. https://doi.org/10.1007/s10676-009-9199-5.

Solomon, J. (1987). New thoughts on teacher education. *Oxford Review of Education*, *13*(3), 267–274.

Sotamaa, O. (2007). Perceptions of player in game design literature. In: B. Akira (ed.) *Situated Play: Proceedings of the 2007 Digital Games Research Association Conference*, (pp. 456–465). The University of Tokyo, Tokyo.

Sotamaa, O., Ermi, L., Jäppinen, A., Laukkanen, T., Mäyrä, F., & Nummela, J. (2005). The Role of Players in Game Design: A Methodological Perspective. In Proceedings of the 6th DAC Conference, (pp. 34 – 42). Copenhagen.

Súilleabháin, G. Ó., & Sime, J. A. (2010). Games for learning and learning transfer. In Roisin Donnelly, K.K.C.O., Jen Harvey (ed.) Critical design and effective tools for e-learning in higher education: Theory into practice, (pp. 113–126). IGI Global Publication.

Ueda, F. (2005). Shadow of the Colussus. Sony Computer Entertainment.

Wilson, D., Sicart, M. (2010). Now it's personal: on abusive game design. In *Proceedings of the International Academic Conference on the Future of Game Design and Technology*, Futureplay '10, (pp. 40–47). ACM, New York, NY, USA. https://doi.org/10.1145/1920778.1920785.

Winn, B. (2008). The design, play, and experience framework. In *Handbook of Research on Effective Electronic Gaming in Education*, (pp. 1010–1024). IGI Global Publication, Hershey, Philadelphia.

Yeh, Y. C. (2004). Nurturing reflective teaching during critical-thinking instruction in a computer simulation program. *Computer & Education*, *42*(2), 181–194. https://doi.org/10.1016/S0360-1315(03)00071-X.

Author Biography

Dr. Rilla Khaled is an Associate Professor of Design and Computation Arts at Concordia University in Montréal, and the director of Concordia University's Technoculture, Art, and Games (TAG) Research Centre. Her research focuses on the design of learning and persuasive games, interactions between games and culture, and methods and practices of game design. Her research and publications on game design have been presented at venues including CHI, GDC, DIS, and INTERACT. Dr. Khaled has worked on a number of large scale game design projects, serving as lead designer on the EU FP7-funded projects SIREN and ILearnRW. Her current research focuses on the ideas of Reflective Game Design and Speculative Play, both of which concern the use of games and play to promote thoughtful interaction and learning.

Playing the Subject

Tom Penney and Florian 'Floyd' Mueller

Abstract Today identities are considered fragmented and multiple; they are ever-changing performances. However, recent discourse surrounding identity suggests the way we engage in online media can actually essentialise identities through social sorting, creating positive feedback loops and by commodifying niche communities. We illustrate our thinking by looking at examples of current online applications that are concerned with identity and investigate how artists play with and subvert these constructs by playing many selves and producing caricatures. We do this in order to advance a discourse of identity in an age of pervasive social media.

1 Introduction

Today we express our identity more and more through networked platforms. The rise of social networking and smartphone applications has greatly increased the speed and frequency at which users can write versions of their identity into being and update them, through systems like Facebook, Twitter, Instagram and Blogger. By using these many services, our expression of identity has become fragmented rather than solid; versions of identity are composed of constantly changing digital performances. The enactment of varied identities online invokes an image of diverse subjects that are formed contextually rather than innately and have no single 'essential' representation. Ironically, however, media systems not only sort and manage identities, essentialising rather than diversifying them, but also create

T. Penney (✉)
Playable Media Lab, RMIT University, La Trobe St,
Melbourne 3000, VIC, Australia
e-mail: thomas.penney@rmit.edu.au

F. 'F'. Mueller
Exertion Games Lab, RMIT University, La Trobe St,
Melbourne 3000, VIC, Australia
e-mail: floyd@exertiongameslab.org

© Springer Nature Singapore Pte Ltd. 2018
D. Cermak-Sassenrath (ed.), *Playful Disruption of Digital Media*,
Gaming Media and Social Effects, https://doi.org/10.1007/978-981-10-1891-6_2

spaces for us to produce, consume and become the stereotypes that we, as users, create. Algorithms can sort and manage the identities of users, create social feedback loops and close off niche communities.

In order to discuss these issues, we turn to art to consider how such constructs might be subverted or played. We look at portraiture and caricature as strategies in the first author, Tom Penney's, own practice as well as in the practices of other artists who have engaged with media and identity historically and currently. These include Cindy Sherman, Candice Breitz, Ed Fornieles and Carla Adams in order to demonstrate how such investigations have been advanced materially and conceptually. We specifically discuss the strategies of playing many selves and the use of caricature, how they function to critique or subvert ideas of an essential subject. Ultimately, although social media can serve to essentialise rather than diversify identities, we want to advocate that it can still be played by artists in order to problematise and critique the effect media constructions can have on identity to advance this discourse in an age of pervasive social media.

2   The Fragmented Subject

The question of identity and the assertion of a 'self' has, and continues to be, an ongoing concern in visual culture (Doy 2005, 1f.). When we speak about an identity, we often refer to its subjectivity. Subjectivity refers to the status of being a subject and expressing its individual beliefs, desires, feelings and impressions. These are what formulate a sense of identity. When we talk about identities we have 'an awareness of what constitutes an individual self, how that self relates to society and the various characteristics that are involved in the construction of subjectivity, such as gender, class, ethnicity [and] sexuality…' (Doy 2005, 6f.). In light of Barthes' *Death of the Author* (1968), the postmodern position on the subject has been to deny its existence. Barthes challenged traditional notions of the unique subject, forwarding that multiple, individual views construct different readings of anything a subject produces. Related positions have since held that all notions of the individual subject, and its identity, are therefore constructed socially or performed. 'Performed' here refers to how people portray different images of 'self' with the knowledge of the different contexts they might be received in. The feminist theorist Judith Butler has done much to affirm this through the lens of gender. In her landmark text *Gender Trouble: Feminism and the Subversion of Identity* (1990), Butler outlines a position on gender that claims all notions of it are performed, rather than 'essential'. If we speak about an 'essential' self, we refer to how an identity can be fixed via searching for its essential core, asking 'what makes you, you?'. This post-structuralist position pushes for gender to be flexible and free floating in the way rigid socially formed categories of 'male' and 'female' do not allow. She argues that essential qualities only appear to exist because of the 'stylised repetition of acts through time' (Butler 1988, 1f.). While this chapter does respond to gender specifically, it more widely concerns the portrayal of identities through social media via acts that are performed. The way we 'write ourselves into being' (Light et al. 2009) through social media is a contemporary example of subjects that are constantly changing and updating their performances in social environments.

3   The Essential Subject

In a 'post-postmodern' world, we are encouraged as creative users to write our subjective voices into social media. Mark Nunes has written about the effects of this on Facebook in *Ecstatic Updates: Facebook, Identity and the Fragmented Self* (2013). In general, capitalist media encourages self-essentialisation for the purpose of marketing; cosmetic surgeons claim the discovery of a 'real you' through their products, and the spirituality industry claims to help people find an 'inner self' through their services (van Zoonen 2013, 49f.). Despite the developments of postmodern feminist theory, the concept of an 'essential self' that resides in individuals is still something that drives consumption. Social media encourages people to express this self; selves affirm their individuality at all costs. On Facebook, this is sold as a kind of 'social good; to contribute to society now becomes an act of contributing content within communicative capitalism's "fantasy of abundance"' (Nunes 2013, 10f.). Mark Zuckerberg, owner of Facebook, speaks of a scenario where 'the world will be better if you share more' (Nunes 2013, 10f.). Finding an 'essential self', a fixed identity, is a marketing strategy; the essential self has become an undeniable focus of our popular culture and media services.

Despite being encouraged to express our subjectivity in social media, we actually participate in others' systems. The degree to which identity 'belongs' to unique individuals is suspect. By 'sharing more' we give the systems interpreting this data more to work with. Within the frame of Facebook, for example, content not only legally belongs to Facebook, but users are understood as a 'vertex or node… [that]… marks an identity' (Nunes 2013, 15f.). The user is not a multiple but rather an aggregate; a whole combined from disparate elements that are the relationships or 'actions and associations' that converge upon the 'vertex' (Nunes 2013, 14f.). The individual is understood through an 'algorithmically generated data profile of contacts and keywords that defines a user as 'dividual' or 'instance' within a larger relational database' (Nunes 2013, 12f.). While we are encouraged to creatively communicate ourselves at great speed, and this behaviour may seem fragmenting, the multiplicity of these performances constitute a whole from the invisible perspective of the digital spaces we subject ourselves to. This leads us to the question of whether identities are defined or limited, despite certain claims for subjectivity; that users can communicate in social networking 'as they so desire' (Nunes 2013, 13f).

In *From Identity to Identification: Fixating the Fragmented Self* (2013), Liesbet van Zoonen expresses a similar perspective. Although cultural theory, especially through queer studies and intersectionality (van Zoonen 2013, 44f.), has tended to consider identity as 'multiple rather than single... dynamic rather than static... and... volatile rather than consistent', van Zoonen discusses social sorting and identity management as phenomena in Internet media that actually impose fixed identities on users. If we take Google's recent change in privacy policy as an example, which collects information from all of its services (Gmail, YouTube, Google+) and merges them into a single account in order to 'provide better services... from figuring out basic stuff like which language you speak, to more complex things like which ads you'll find most useful or the people who matter most to you online' (van Zoonen 2013, 47f.), we gain a picture of how services are defining us as specific 'types' of user. The service is sold as supporting individual subjects through a user-friendly interface that 'gives you what you want'; however, this data is used by Google, in a similar way to Facebook, to define a user in a network. Using this definition, Google then displays only certain types of content to you. Social sorting denies users the possibility of a multiple self, says van Zoonen. Out of 'necessity, it needs to identify people as belonging either in one or the other category, but definitely not in more than one' (van Zoonen 2013, 47f.). Van Zoonen acknowledges 'diversity' as a desirable goal for social and cultural policy but questions whether this is being achieved. Google can serve to render this 'diversity' through essential qualities, a certain 'type' of diversity.

Social networking systems may sort or manage us, essentialising a diverse or multiple self, but we willingly produce the content that constructs the 'types' they identify for us. We are encouraged to express 'essential selves' through social media, but then social media, having used this data, reflects its own essentialisation back. In Baudrillard's terms (1988, 12f.) 'The scene and the mirror have given way to a screen and a network', but the Internet is starting to look like a mirror that tells us what we already are, what we want to hear, rather than a mirror that silently reflects. By using an interface that purports to give us 'what we want', we may have thought we were telling the Internet what we want to see by filling out questionnaires, endorsing content, adding certain friends or check-marking different boxes. The virtual image of the self that synthesises an essential 'type' from all the data we input is something we consume reflexively. Through the screen-as-mirror, a positive feedback loop, are we thus consuming the Internet's caricatures, its 'extreme versions', of ourselves? A manifestation of this concept is the echo chamber, which applies more specifically to blogs and blog culture. By literally writing ourselves into cyberspace, an echo chamber can been invoked; 'a condition arising in an online community where participants find their own opinion constantly "echoed" back to them, thus reinforcing a certain sense of truth that resonates with their individual belief systems'. (McCrae 2011, 1f.). Not only does the *echo chamber* create a closed circle of feedback for individuals, but it can limit entire communities to the circulation of closed opinions, acting to shut down effective critical discourse (McCrae 2011, 3f.).

4   Commodifying Difference

Essentialised subjects and feedback loops have implications for supporters of diverse identity, especially given Butler's assertion of a fluid concept of sexuality. We believe this is especially true of online dating applications which encourage users, in many cases, to limit both themselves and what they look for in other people, to rigid 'types'. The websites and applications in question are often specifically tailored to minorities, but encourage users to subscribe to particular images of those minorities. Major contributors to the development of such 'types' are indeed the users themselves who perpetuate certain mannerisms, languages and looks through their collective use of those systems. These in turn become the 'repeated acts' invoked by Butler than can serve to cosmetically re-essentialise certain forms of identity. Here, we use the extreme case of Grindr, which is a mobile dating application for gay men with highly specific codes of engagement. It presents users with a library of faces or bodies that can be browsed through before conversation is initiated. It uses a GPS system to detect which users are closest to you. The application is commonly known to be used for sexual 'hook-ups' although it can be used for casual dating or meeting friends. Opening Grindr, one is visually presented with an array of flat, narcissistic 'selfies'. These are photographs taken on a phone of oneself, often at arm's length, directed at the face, or using a bathroom mirror. The app presents a grid of squares containing these selfies where all subjects are equalised through their visual representation under the surface of the mobile phone. Swiping through the profiles-under-glass is like flicking through an IKEA catalogue; these bodies are defined as potential objects for sale.

Ben Light has researched the construction of homosexual identity on a comparable system, Gaydar.co.uk, noting that users of it subscribe to versions of homosexuality that may contribute to 'marginalisation in society… through the deployment of strategies based on the commodification of difference' (Light et al. 2009). The commodification of difference here refers to the way a niche market has been identified and turned into a product, a consumer choice. Marginalisation through this commodity is indeed perpetuated. Codes of communication are highly specific to the gay Grindr community. Users define themselves through codes like 'sub' (submissive), 'top' (dominant), 'GAM' (gay Asian male). They have control over what they do or do not want the system to provide, by limiting the age range of searches, for example, or by making their desires explicit through statements on their profile, such as 'I only want fun', which in the Grindr universe translates to 'I only want to have sex'. As a result of niche marketing, 'Such conceptions of gayness may be stereotypical and defined against a heterosexual norm, rather than intersecting with complex identities that include multitudinous forms of gender and sexuality' (Light et al. 2009). As a user one feels an enormous amount of pressure to appeal to the stereotypes and codes that are perpetuated and constantly written by users into the 'community' that Grindr is:

> Through the use of Gaydar.co.uk [and Grindr] individuals write a version of themselves and of this gay community into being. However, because of the desire to commodify 'the

difference' that is gay, predominantly white men, online and offline, such inscriptions become monolithic caricatures that are obdurate and enrol even those who do not participate in such arrangements at all or only by proxy. (Light et al. 2009)

On Grindr users can participate in these inscriptions by taking selfies of themselves and using them on their public profiles. Lacanian thought tells us that identities are formed when children recognise their reflections in the mirror; the mirror's representation of the body allows the body to be perceived as an object that can be compared against other objects of language (Lacan 1949, 503f.). If '…media images act as Lacanian mirrors that cause identity formations to be ideologically laden' (Peretti 1996), then Grindr allows users to participate as a media image—an object participating in this language. Users' images on the screen (their selfies) are displayed alongside other profiles on Grindr as squares comparable to others around them. Peretti (1996) further asserts that, 'In Lacanian terms, consumer capitalism needs subjects who continually re-enact the infantile drama of mirror stage identifications.' in order to maintain productivity. The app rapidly updates profiles and refreshes the indication of proximity they have to each other leading subjects to constantly question their portrayal as a media object. This is not a seasonal change of fashions but a minute-by-minute update of images. As soon as external selves are consolidated by single pictures, they are immediately threatened again when compared against to a context of other profile images. By participating this way, through the narcissistic portrayal of many selves, one feeds Grindr the images that construct its types through a positive feedback loop, a mirror.

5   Artistic Responses

We have raised a number of concerns for identity, or the subject, that artists might respond to or use as a point of departure for creative work, particularly, the notion of a fluid or multiple self as something that can be limited and essentialised through Facebook, Google or Grindr. The multiple self is of great importance to artists who have long romanticised their notion of self resistant and irreducible to definition by others' systems and stereotypes. We propose that artists can visually and conceptually analyse what it means for social media systems to essentialise selves in the contemporary world, and subvert this in the process. Many artists have invoked 'acts' or 'performances' in new media as a way to question 'who is playing who?'. The disappearance of artists into systems through these performances, however, makes it difficult to locate a critique, and at points, artists can be seen as complicit within them. We therefore consider how the use of caricatures of individuals as well as of whole systems can function to critique the discussed systems.

6 Playing Many Selves

Artists like Cindy Sherman have played many selves in order to subvert media representations of women. In her work *The Untitled Film Stills* (1977–80), Sherman dressed up and photographed herself as different 'types' of women in popular film, television and printed media. Of her work Sherman states 'I feel I'm anonymous in my work... "When I look at the pictures, I never see myself; they aren't self-portraits. Sometimes I disappear"' (Collins 1990). Part of what we see in the work of artists that play many selves is that they 'disappear' into a multiplicity of performances through their mimicry of external acts and media portrayals. This is a key strategy artists use to disrupt the relationship between media and the subject in order to highlight or draw attention to it. The history of artists probing identity through media has been recently presented in Candice Breitzx's *The Character* (2013) at the Australian Centre for the Moving Image (ACMI). Candice Breitz's videos in particular ask 'To what extent are our lives "scripted" for us by the media we consume and other influences that we encounter in our intimate and social environments?' (ACMI 2013).

In Breitz's fourteen channel video installation, *Becoming* (2003), Breitz faithfully re-enacted short segments of female performances in Hollywood films. Her own video performances are shown on standing screens with footage playing on either side; one side shows the original Hollywood footage in colour and on the reverse side is Breitz's re-enactment in black and white. Breitz mimics performances, for example, by Julia Roberts in *Pretty Woman* and Cameron Diaz in *The Sweetest Thing* with uncanny accuracy and exacting body language. The nature of the dual footage, which is synced exactly and of precisely the same duration, makes it indiscernible which person, the artist or the actor, is speaking the dialogue. The nature of Breitz's portrayal of the actresses, if not necessarily critiquing them, constructs a scenario where viewers ask themselves 'who is playing who?'. This occurs in the moment we switch from one side of each pair of videos to the other in astonishment as we attempt to figure out who is speaking the dialogue. Breitz's work acts to decentre her identity and the identity of other subjects by confusing the boundary between the single actor and the many roles. This is especially true in her video installations '*Him*' (1968–2008) and '*Her*' (1978–2008) which feature multiple videos of Jack Nicholson and Meryl Streep, respectively. In each video, multiple sections of films the actors have played in are extracted and are played, cut together simultaneously, with each other. Background in each video has been removed so that we only see the actors. It is difficult to tell whether through the many different 'acts' the images collectively form an essential picture of the actors, or whether at the end of the day, it should be accepted that none of those images are 'the actors' because all are simply performed roles. This does not matter however, as the contrast between the two thoughts raises awareness of this disappearance. As such, this disappearance is a point of focus getting viewers to question 'who is playing who' out of confusion.

Artists advocating the play of many selves not only disappear into the roles they portray, but can disappear into the systems that produce them as well. Ed Fornieles investigates many selves through their disappearance into such systems. He is a digital 'post-Internet' artist who has used social media to examine the 'types' and identities inhabiting it from within. In *Dorm Daze* (2011) also called The Facebook Sitcom, 'participants inhabited profiles scalped from real life American college students' (McNeil 2012). A feature of Fornieles' work is that individuals play other individuals, not celebrities. Individuals are written as larger-than-life types online; they are the 'celebrities'. For *Dorm Daze,* the result was a series of narratives formed from the evolving discussions between characters. One can read the conversations between members of the fictional *Orca* or *Sigma Chi* houses, and the various political and religious stereotypes they come to embody through their exchanges. While the fiction plays out through Facebook itself, it has informed wider projects and strings of thought. *Animal House* (2011) was a 'series of college party performances which emulated the dorm/frat environment' (McNeil 2012) where over 200 people had been assigned roles for what 'type' of college character they would play throughout the event. Projects such as these are opportunities to consider the identity of communities, which essentially form stereotypes, and how easily the participants come to embody these by naturally adopting the codes, modes of speech and mannerisms that define the roles they have been given. By getting many participants to play many other participants, Fornieles can get them, by quickly adopting various roles, to consider their ability to become multiple or other selves. For our discussion this means that by playing each other, we are encouraged to disengage from our regular feedback loops and consider interfacing with a system from a different point of view, a strategy that decentres the essentialised self.

Playing multiple selves in digital spaces however complicates the multiple/essential binary. New media artist Mark Amerika champions this kind of production which he dubs the 'one person art-making machine' (Amerika 2008, 77f.). The 'digital-artist-to-be' must constantly 'play themselves—even if that means having to reinvent their artistic personas over and over again' (Amerika 2008, 82f.). Amerika praises a sort of 'becoming one' with the system while maintaining a nomadic fluidity within it. Amerika speaks of 'remixologists—performance artists who manipulate all of the useful data they have sampled from so that they can then reconfigure their own stories into a pseudoautobiographical narrative that spins the media attention right their way' (Amerika 2008, 77f.). An emphasis on the artist 'selling themselves' is prevalent. While this voyage may be free-form, flux or nomadic, it does of course still tiptoe around the rhetoric of 'selling oneself' or 'being who one wants to be'. Larissa Hjorth (2013a, 100f.) says 'some artists are productively using Facebook to send out invitations, others are using it to perform a type of public intimacy in which messages, photographs and newsfeeds all catalogue and cultivate the image (and aura) of the artists'. Such a mode of artistic productivity is increasing with the level of comfort younger artists have in using technology. In a study of artists born after 1989 ('89+'), Hans Ulrich Obrist has referred to 'the diamond generation', a 'tagline for [a] group of fluidly networked and distributed digital natives' (Burke 2013). The 'digital natives' 'develop [their] practices in tandem with the very economies and structures that maintain them, and in fact face these structures at their most blatant and abrasive. Artistic agency lies in the degree to which these structures are incorporated and subverted'. (Burke 2013) This suggests that the act of 'disappearance' will only increase as generations become more implicit within social media, and artistic outcomes will blur more and more with the content of 'non-artists'. In this disappearance, we must ask ourselves whether artists are therefore still pushing for the playing of many selves in order to subvert systems, encouraging multiplicity, or whether this is misleading; that rather, many selves are the way artists realise their 'essence' in a multiplicity of outcomes and become complicit within the systems that frame them.

7   The Fractal Subject

It seems that although artists fragment themselves and others within systems, their subjects are still paramount to framing the work. Something confusing about the union of artists, their roles and the systems they play in new media is their politics. Are they 'for' or 'against' the systems they become one with, and how can we tell them apart? I addressed this irony in a project of mine; *Everyone's a Hero in Valhalla* (2010) by responding to the words of new media artist Roy Ascot; 'I long to be many selves distributed through the networks... I AM many selves... I wish to recognise my potential of being immortal in that sense' (Ascot 1993). In his words, I see an ironic construction; through his many selves he wishes to realise an *immortal self*; it seems quite an egocentric concept that downplays his desire to be multiple. Jean Baudrillard has described *The Fractal Subject*, something that here positions the irony of a disappearing, many-self-playing artist:

> ...one can speak of the fractal subject... diffracted into a multitude of identical miniaturized egos... completely saturating its environment... the fractal subject dreams only of resembling himself [sic] in each one of his fractions. That is to say, his [sic] dream involutes below all representation towards the smallest molecular fraction of himself [sic]; a strange Narcissus, no longer dreaming of his ideal image, but of a formula to genetically reproduce himself [sic] into infinity (Baudrillard 1988).

The idea of a fractal subject is influential to my work. In terms of our discussion, this is a subject who is neither multiple, nor singular and essential. As both, it is a multiple that realises its essence in many outcomes. I had seen elements of the fractal subject within the use of Facebook, where 'versions' of self are curated in libraries of profile images. When browsed, one can realise their essence within each performance despite their multiplicity and difference. One's identity is not composed out of an aggregate of many selves, but rather realises itself in many agents. When thinking about the fractal subject, I liken its multiple agents to '*horcruxes*' (from J.K. Rowling's *Harry Potter* (1997–2007)) which are a set of objects (seven in Rowling's book) containing fragments of a soul that has been split between them.

The subject whose soul has been split will never die so long as one horcrux remains intact. In Roy Ascot's sense, one becomes immortal. In 2009, I had made a series of artworks called *The Horcruxes*, where I 3D-scanned various artworks of mine and turned them into digital images. In the Duchampian tradition, I signed each object using Photoshop (inscribing my 'essence' within each) and made them available online. I asked people to download them as much as possible; the more downloads of each horcrux, the more my soul would divide. The more pieces of my soul that existed, the more difficult I would be to kill. This was my allegorical playing of the fractal subject. In this example, my strategy was to embody the fractal subject rather than oppose it.

I then attempted to subvert the *fractal subject* by creating *The Tarot Self Portrait* (2010) (Fig. 1). I posed as all 72 cards of the Tarot. The images were printed as actual cards and contained in a self-branded box; '*Tom Penney is the Tarot*'. I wanted to ask a question; if I subject an entire system of human reflection to my own individual body and if I appropriated each of its archetypes, would I therefore represent a subject that not only narcissistically imagined itself as immortal through its multiplicity, but through whom others would be forced to express if they played the cards? I elevated myself to the same system and standard as these universal archetypes. In essence, I imagined a system whereby if I played the cards, I would be tautologically consuming and reflecting upon only versions of myself, and if others reflected upon the cards, they would understand themselves only in terms of my system. Both the archetypes and whoever uses the cards become expressions of my *fractal subject* not their own. In *The Tarot Self Portrait*, however, none of the performances are me. They are images separate to my body (as are one's images on Facebook). Wanting to represent this irony within the work, I played each role through cheap tacky costumes and the poses appear sarcastic unlike the seriousness

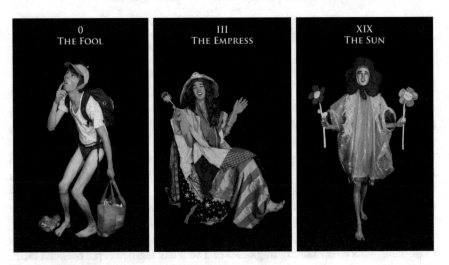

Fig. 1 Tom Penney (2010). Images from The Tarot self portrait

and universality of the archetypes they represent. My expressions and poses are absurd, humorous and playful, forming irreverent caricatures of the Tarot archetypes. This strategy serves to clearly reveal a performance rather than disappear into one. There is no question of 'who is playing who?' My playfulness manifesting in caricature becomes a strategy to expose irony, rather than disappear into, a system and artistic position that encourages individuals to participate as fractal and narcissistic.

At this point, we realise that aesthetic strategies examining the play of many selves today can be critical rather than implicit. My work, *The New Spiritual Network* (2010), consists of nineteen videos depicting me in various farcical roles. I portrayed witches, gurus, academics and cultists all peddling their own subjective methods for finding an 'inner self'. My intention for *The New Spiritual Network* was to render absurd the relationship between 'essential selves' sought through different spiritual perspectives and their reality as a subjective multiplicity. Many individuals sell their 'sure fire methods' for spiritual journeying through YouTube. In one set of videos, I had used six people to listen to a famous speech on YouTube ('*Re:Evolution*') by new age shamanic guru and theorist, Terence McKenna. Each performer had to repeat his words received via headphones immediately and without pause or rehearsal. I had intended for the effect to mirror a 'channelling' of a spirit through one's body and speech as a psychic might do. The new age spirituality babble is stuttered and inaccurately delivered by each performer as they receive the sound via a YouTube video. The matching of the performer to the 'influencing media' is broken. Brokenness serves not only to make viewers question the origin of the subjective words ('who is playing who?'). One questions whether the performers are in fact complicit in speaking them ('are they happy to be played?'). This disunity rather serves to critique than question; it suggests the performers cannot handle and are controlled by, rather than simply reflect or 'disappear into' the digitally provided speech.

8 Caricature

In each of my works, caricature becomes an important strategy of critique. The absurd clumsiness, reduction and brokenness of my artworks emphasise questionable aspects of relationships between subjects and media. They are caricatures of systems as much as they are of the subjects operating in them. Here, I emphasise caricature as a strategy to critique social media. In *Notes on Caricature* (1989), Mike Kelley discusses the function of caricature. As a kind of antithesis to modern abstraction, caricature uses reductive processes to achieve 'a portrait that deliberately transforms the features of its victims so as to expose and exaggerate their faults and weaknesses' (Kelley 1989, 20f.). Kelley points out that 'although they may appear to be very different, caricature, which uses deformation in the service of ridicule, and the idealised, heroic, classical portrait, are founded in similar essentialist assumptions' (Kelley 1989, 22f.). The key here is that the purpose of caricature is to critique, not immortalise or affirm the essential qualities that portraiture attempts to draw out. It does this by highlighting certain features of the critiqued. At many points, I have alluded to our engagement with the screen-as-mirror as defining 'types' of users by synthesising personal data into an essential 'version' of a user, not unlike caricature. Social media, however, does not choose to do this for the purpose of critique. It rather synthesises information based on data about individuals in order to optimise an understanding of one and streamline their networked experience. The creative artist, however, has every opportunity to subjectively represent faults for the purposes of critique through caricature. As a strategy, this serves to distance the artist critically from the critiqued by subjecting subjects and systems to their own play.

A playful approach to caricature was exemplified in *The One Minute Soul Capture* (2009) (Fig. 2), an artwork of mine in response to the computer's reductive role in essentialising individuals. This project was framed through my performance as a witch who having lost her magic powers had to use computers to perform dark magic ('*The dark arts of art*'). She seized the profile images of my Facebook friends and digitally transformed them, each in the space of one minute, using quick and cheap filters on Photoshop. This process was filmed sarcastically as an informational crafts show, where the witch showed you how to do it at home using your own Photoshop software. Each subject was reduced to an unflattering digital caricature. Similar to *The Horcruxes,* its outcomes were essentialisations of individuals realised in digital artefacts; however, they were negative depictions rather than celebratory. While this could be interpreted as a critique of the narcissism of each subject's self-representation on Facebook, I had masked their identities through each image and as a collection the work became more about the overall process of reduction. The gesture of 'the filter' became metaphoric for the subjective essentialisation a computer perceives on its end of the screen-as-mirror. The framing of this reduction as dark magic placed it in a critical light by rendering it an 'evil' act. With a screeching Python-esque voice,

Fig. 2 Tom Penney (2010). Images from The One Minute Soul Capture

shoddy makeup, and wearing a torn sheet and crocs, the character of the witch who performed the dark magic became a meta-caricature. As a witch who conflated digital processes and dark magic, who 'became one with the filter', she was a caricature of systems producing caricatures of subjects. The witch subjected others to her own system; the *soul captures* became agents of her expression, and the individuals in them had no agency in their representation. A fractal subject was invoked yet again, but this was a caricature of one. I critically distanced myself from the fractal subject by here performing in a role that was not 'me'. My aim was not to disappear. By playing roles-as-caricatures and subjects-as-caricatures, my intention is to critique.

These examples exemplify how we see caricature as a strategy that places control in the hands of the artist. Artists can subjectively manipulate and re-imagine their subjects through reduction, distortion and exaggeration. For me, playing with caricature is a way of taking back dominance over systems that essentialise us. Mary Flanagan through her body of work on *Critical Play* (2008) discusses how artists, non-artists and children have played with dolls and doll houses as a way of understanding, controlling and subverting the systems of domesticity and self-image they represent. One can symbolise whole systems and the characters playing within them as a way to control them. In computer gaming, this has been achieved through computer games like *The Sims* (Maxis 2000–current) or *Sim City* (Maxis 1989–2013), which give players virtual worlds to manipulate but also offer them opportunities to arrange those worlds as they wish; one can deviate from the expected path to a great degree. Playing with bodies and worlds in the form of virtual and physical representations allows for a kind of '*transformative play*', where one can overcome a structure in order to imagine it differently (Salen and Zimmerman 2004, 305f.).

Artist Carla Adams and I treat our subjects as dolls subjected to a process of caricature. This allows us to take control over the subject's representation in a virtual environment. We draw over subjects and emphasise different qualities in order to portray their weaknesses in a digital domestic system. This domestic system, for both of us, has largely involved online dating. Our *transformations* are caricatures that imagine structures as broken and forlorn rather than functioning. Much of Adams' recent works have involved her encounters with men through *Omegle*; she takes images of the people she interacts with via webcam and paints over them so as to mask their identity but emphasise their flawed nature. In *Very Sad Men* (2012) (Fig. 3), the subjects are anonymous but the nature of the colour

Fig. 3 Carla Adams (2012). Images from Very Sad Men

and shape in each image is blob-like, crude and unflattering. The colours are sickly or pale. These caricatures emphasise sadness in each subject from the subjective point of view of the artist, being the only feature that sings through each. Adams has also recently constructed her caricatures in sculptural form, using papier maché to create lumpy, sagging versions of her online encounters. The effect of these caricatures is to render them powerless as objects on the other side of a screen, reversing a relationship that would normally position the female as the object of a male gaze. In this scenario, the men become objects as artworks that can be positioned or played with like dolls in order to subvert the new domesticity of social media. By imagining this environment differently through caricature, the sadness of anonymous webcam interaction is emphasised and critiqued by the artist.

Building on this, I expand caricature as control to also include the environments, the 'doll houses' that such caricatures operate in. Recently, I have created a series of images based on *Grindr* that are designed to be viewed by swiping through a self-designed app. I use the popular new computer game-building software *Unity3D* to achieve these caricatures and build the interactive environment in which they appear. *Unity3D* presents me with a three-dimensional plane where objects can be arranged, built, scaled, reduced and skinned. By dropping objects into the three-dimensional interface, I can imagine whole environments where objects interrelate in a virtual space and produce relationships that construct meaning, not unlike the visual arrangement of elements in a painting. When I play around with my 3D images, I feel like I am controlling a kind of computer game world; I am used to playing simulation games where one controls different characters, popu-lations and environments. My *Unity3D* worlds become symbolic doll houses, here ones that are caricatures of the systems that subjects might operate within. My self-designed app that mimics *Grindr* has been made this way. I had created a series of male 'dolls' in *The Sims 3* each with their own personalities, looks and traits. I then used these as the origin for models that I placed in a *Unity3D* world. I have imperfectly used a separate 3D scanning application to reduce the 'dolls' to inaccurate, painterly reductions that erase the individuality of each character, before adding a perfectly modelled iPhone back into their hands; the caricatures are taking 'selfies'. The images (Fig. 4) have been designed to reflect *Grindr* profiles by their placement into an orange frame and a system that organises them to be viewed through a smartphone. Using only the characteristic orange of the *Grindr* interface, the same resolution of the smartphone screen and its swiping gesture, the app reduces the features of *Grindr* to a symbolic level and acts as its critical double.

The use of caricature to represent whole interactive environments sets artists apart from media systems rather than see them operate from within and disappear. This kind of caricature, system-as-caricature, not only has implications for artists but for designers of computer games and interactive media. *Machinima*, and *art games*, for example, has combined computer games, visual art and cinema in the Dadaist tradition of playful subversion (Hjorth 2013a, b). In the spirit of Flanagan's *Critical Play* (2009) knowledge of games, art and new media can combine to create critical creative outcomes that interrogate our methods of living through social media. A method of constructing a caricature of a system is to use basic gestures of

Fig. 4 Tom Penney (2013). Images from Selfies

interaction, such as the swiping of the finger on my *Untitled* Grindr app, as symbolic for 'interaction'. In a recent work also made in *Unity3D*, *Fragile Ego* (2013) (Fig. 5), I have exaggerated the feature of the 'like' button on Facebook. By reducing the Facebook environment to two symbolic actions (clicking a 'like' or 'dislike' button), viewers are able to inflate or deflate phallic monster characters contained within a box reminiscent of a Facebook page until they explode. This gesture acts to form a critical representation of Facebook by emphasising the

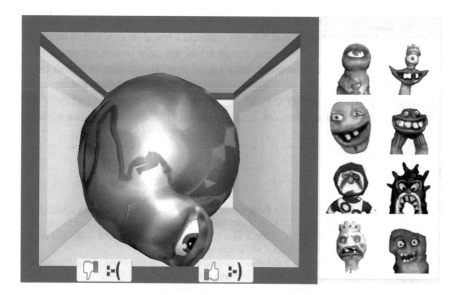

Fig. 5 Tom Penney (2013). Screenshot of Fragile Ego

relationship between 'input' and 'ego' (more clicking = bigger ego, less click-ing = deflated ego), highlighting this functionality by rendering it simplified and absurd. An entire system is criticised through two buttons. This simple system acts as a caricature of Facebook through the emphasis of a single mode of interaction.

9 Conclusion

In summary, through an analysis of existing works and my own, we have identified a number of strategies that subvert the essentialisation of identity in online systems. We first discussed the idea of an 'essential self' as something undermined by postmodern theory, but then noted that social media today encourages self-essentialisation through the examples of identity management, positive feed-back loops and commodification of niche communities. The first strategy we identified as subverting essential selves has been to play many selves in order to represent the subject as multiple and intersecting with other 'types', celebrities or other users, in popular and social media. While this has decentred the subject by asking 'who is playing who?' it can lead to a situation where artists politically disappear into the systems they attempt to critique. Given the complacency of digital natives within social media, it is likely that this disappearance will only increase. As social media encourages fractal subjects to realise their essence within many agents, artists, too can come to represent the fractal subject through their many selves and outcomes. To provide a critical alternative to this scenario, we have identified through my own practice, caricature as a way to subversively rep-resent both the subjects of social media and the systems of social media. The effect of doing so positions the artist as separate to the systems under scrutiny by paro-dying social media constructs and the way subjects form identities within them. Our strategies hope to shed light on new media art as not only providing novel strate-gies, but providing critical ones that truly examine what it means to reflect on life through the screen-as-mirror. Without highlighting the changing nature of the subject through contemporary critical art, we fear the total complacency of new generations within systems that can come to define, limit and essentialise diversity.

References

Amerika, M. (2008). Making space for the artist. In M. Alexenberg (Ed.), *Educating Artists for the Future* (pp. 75–82). Chicago, USA: Intellect books, The University of Chicago Press.

Ascot, R. (1993), *Paris 23/10/1993 Part 1 of 12*, Retrieved June 02, 2013 from http://www.youtube.com/watch?v=_ueEeDmoQ7o&feature=related.

Barthes, R. (1968). The Death of the Author. *Roland Barthes: Image Music Text*, translated by Stephen Heath (pp. 142–149). London: Fontana Press.

Baudrillard, J. (1988). *The Ecstasy of Communication*, translated by B. C. Schutze (p. 12). New York: Semiotexte.

Breitz, C. (2003). *Becoming*. Film installation. Melbourne: Australian Centre for the Moving Image.

Breitz, C. (1968–2008). *Him*. Film installation. Melbourne: Australian Centre for the Moving Image.

Breitz, C. (1978–2008), *Her*. Film installation. Melbourne: Australian Centre for the Moving Image.

Burke, H. (2013). Yonger than Rihanna. *Rhizome*. Retrieved February 07, 2018 from http://rhizome.org/editorial/2013/feb/14/younger-rihanna/.

Butler, B. (1988). *Performative acts and gender constitution: An essay in phenomenology and feminist theory,* (p. 1). Retrieved June 01, 2013, from http://www.egs.edu/faculty/judith-butler/biography/.

Butler, J. (1990). *Gender Trouble*. Routledge: New York.

Collins, G. (1990). *A Portraitist's Romp Through Art History*, New York Times, (p. 1). Retrieved June 01, 2013 from http://www.nytimes.com/1990/02/01/arts/a-portraitist-s-romp-through-art-history.html.

Doy, G. (2005). *Picturing the Self: Changing Views of the Subject in Visual Culture* (pp. 1–6). London: J.B Taurus.

Flannagan, M. (2009). *Critical Play: Radical Game Design*. Massachusetts: Massachusetts Institute of Technology Press.

Fornieles, E. (2011). *Animal House*. Performance artwork. London: Guest Projects.

Fornieles, E. (2012). *Dorm Daze*, online Facebook performance. Facebook.

Hjorth, L. (2013a). Frames of Discontent: Social Media, Mobile Intimacy and the Boundaries of Media Practice. In W. Macgregor & H. Koskela (Eds.), *New Visualities, New Technologies: The New Ecstasy of Communication* (pp. 99–118). Burlington USA: Ashgate Publishing.

Hjorth, L. (2013b). The art of games: Machinima and the limits of art games. In J. Ng (Ed.), *Understanding Machinima: Essays on Filmmaking in Virtual Worlds*. Bloomsbury Academic.

Kelley, M. (1989). Foul Perfection: Thoughts on Caricature. In M. Kelley 2003 (Ed.), *Foul Perfection: Essays and Criticism* (pp. 20–22). Massachusetts Institute of Technology.

Lacan, J. (1949). The mirror stage as formative of the function of the I as revealed in psychoanalytic experience. In V. B. Leitch et al. 1997 (Eds.), *The Norton Anthology of Theory and Criticism*. W.W. Norton & Company.

Light, B., Fletcher, G., & Adam, A. E. (2009). Gay men, Gaydar and the commodification of difference. *Information Technology and People, 21*(3), 1–3.

Maxis (2000–current). *The Sims*. Various platforms. Electronic Arts.

Maxis (1989–2013). *Sim City*. PC Computer Game. Electronic Arts.

McCrae, P. (2011). *Echo Chambers and Positive Feedback Loops: The Complex Nature of Echoing Voices on the Internet*, Phil McCrae, (pp. 1–3). Retrieved June 01, 2013, from http://philmcrae.com/uploads/2/8/6/8/2868622/argumentum_ad_infinitum_-_echoing_voices_mcrae.pdf.

McNeil, J. (2012). *Artist Profile: Ed Fornieles*, Rhizome, Retrieved April 15, 2013, from http://rhizome.org/editorial/2012/apr/2/artist-profile-ed-fornieles/.

Nunes, M. (2013). Facebook, identity and the fractal subject. In J. MacGregor & H. Koskela (Eds.), *New Visualities, New Technologies: The New Ecstasy of Communication*, (pp. 7–25). USA: Ashgate Publishing, Burlington.

Peretti, J (1996). *Capitalism and Schizophrenia Contemporary Visual Culture and the Acceleration of Identity Formation/Dissolution*, Negations, Retrieved June 01, 2013, from http://www.datawranglers.com/negations/issues/96w/96w_peretti.html.

Salen, K., & Zimmerman, E. (2004). *Rules of Play* (p. 305). Massachusetts: Massachusetts Institute of Technology Press.

Rowling, J. (1997–2007). *Harry Potter*. London: Bloomsbury.

Sherman, C. (1977–1980). *Untitled Film Stills*, 69 black and white photographs. New York: Museum of Modern Art.

Unknown (2013). *Candice Breitz: The Character*, ACMI, Retrieved June 01, 2013, from http://www.acmi.net.au/candice-breitz.aspx.
Van Zoonen, L. (2013). From identity to identification: Fixating the fragmented self. *Media, Culture and Society, 35*(1), 44–49.

Author Biographies

Tom Penney is a lecturer in Digital Media and Games at RMIT University and was previously the Industry Fellow of Digital Media in that program. His practice and teaching involves 3D imaging, games technology and digital design as well as traditional art media. He has shown work and published through exhibitions, journals and conferences including Media International Australia, The Feminist Journal of Art and Digital Culture, The International Journal of Contemporary Humanities, Perth Institute of Contemporary Art, West Space, The Substation, Laznia Arts Centre, Federation Square, Metro Arts and the International Symposium of Electronic Art. His PhD research project "Critical Affection" investigated an expanded notion of "critical play" and affect through digital contemporary art practice. Much of this drew on a series of interactive 3D artworks that critically represented online dating apps. Narcissism, affection-images, intimacy and the consumption of bodies-as-objects through such apps were critiqued through Tom's use of digital caricature and presentation of virtual, manipulable bodies in interactive works. Tom previously worked as a project manager and artist at Metaverse Makeovers, an augmented reality fashion company, as well as teaching art and design at Curtin and Monash Universities.

Florian "Floyd" Mueller directs the Exertion Games Lab at RMIT University in Melbourne, Australia. The Exertion Games Lab investigates the design of exertion games; these are digital games that require physical effort. This research is situated within a broader interaction design agenda that supports people's values such as an active and healthy life. Floyd has most recently been a Fulbright Visiting Scholar at Stanford University, having worked on the topic of exertion games now across four continents, including at organizations such as the MIT Media Lab, Media Lab Europe, Fuji-Xerox Palo Alto Laboratories, and Xerox Parc. Floyd has also been a Microsoft Research Asia Fellow and has worked at the Microsoft Beijing lab with the research teams developing Xbox Kinect. Previously in Australia, he has worked at the University of Melbourne and was a principal scientist at the Australian Commonwealth Scientific and Industrial Research Organization (CSIRO), where he led the Connecting People team of 12 researchers

The Potential of the Contradictory in Digital Media—The Example of the Political Art Game *PoliShot*

Susanne Grabowski and Daniel Cermak-Sassenrath

Abstract When we created *PoliShot*, a political Dada game and interactive installation, we were confronted quite unexpectedly with the question of what is morally or ethically tolerable in digital games. When it was exhibited, it provoked shocked and concerned reactions from curators and visitors alike. The stumbling block was the use of violence, or more specifically, asking the players to act violently in the game. We take our experiences as an occasion to enquire into and discuss the contradictions of the actual and the virtual, of concept and content. We attempt to draw historical and contemporary parallels and reflect on how art production is not limited to the work, but includes the artists and the audience as essential players in a dynamic system of meanings, motives, and interpretations, full of (un)intended and (un)anticipated conflicts, provocations, breakdowns and shifts, creating exciting and challenging opportunities for play.

1 Introduction

Despite questionable content, shooting games of all kinds continue to be extremely popular (see, e.g., Fritz and Fehr 2003; JIM-Study 2012; Kolokythas 2013).[1] Roughly speaking, these games reward the player with a high score for skillfully

---

[1]Fritz and Fehr (2003) find that games with violent, aggressive, and warlike content are the most popular ones. The first-person shooter *Call of Duty* is, according to the German JIM-Study (2012:49) one of the favorite games of 12- to 19-year-old boys. At present, eight of the games in the top-ten list of *PC-World* (2013) are shooting games.

S. Grabowski (✉)
compArt: Center of Excellence Digital Art, University of Bremen, Bremen, Germany
e-mail: grabowski@uni-bremen.de

D. Cermak-Sassenrath
ITU Copenhagen, Copenhagen, Denmark
e-mail: dace@itu.dk; mail@dace.de

© Springer Nature Singapore Pte Ltd. 2018
D. Cermak-Sassenrath (ed.), *Playful Disruption of Digital Media*,
Gaming Media and Social Effects, https://doi.org/10.1007/978-981-10-1891-6_3

shooting enemies of some description. For Fritz (1995:23), the games realize the players' disposition towards speed, aggression, instant reward, and action.[2] Murray (1997:146) observes that "fighting game[s]" have technically developed a "tight visceral match between the game controller and the screen action" which affords the player a very direct "sense of agency" and "requires very little imaginative effort." Although the settings of the games are overwhelmingly violent scenarios, only few people appear to refuse outright to play them, that is, reject the act of shooting. Fritz and Fehr (2003:53) see the reasons in the players' wish to realize power, dominance, and control (which are closely linked to violence) often articulated as the injuring of an opponent (suffering is, however, not part of these games). Many players find the games fascinating because they can relate their life situations to the patterns in the games (ibid.:51). If the playing of violent games were motivated by the need or possibility to act out safely, to release or to channel aggressive and destructive impulses, if the games had a compensating and regulating function,[3] they would presumably be accepted by society as useful and valuable tools.

But this is not the case. On the contrary, the games are suspected of having adverse effects on their players and on society in general. They raise fears about connections between violence in games and violent behavior in everyday life, that is, whether one is motivated by the other.[4] Much attention is regularly directed towards the issue after so-called school shootings, such as Littleton (US, 1999) and Erfurt (FRG, 2002). However, Kunczik and Zipfel report in their study (2010) no *significant* correlation between violent actions in play and violent actions in everyday life. Although their findings indicate that it is *possible* that medial violence influences the recipients' aggression levels, the effect is only moderate and temporary (Kunczik and Zipfel 2010:13). Violent medial representations are also only one factor in a complex network of reasons and causes for the occurrence of physical violence, and computer games have no greater impact than other media (ibid.).

What is striking about violent game scenarios—with the exception of military games about the most popular wars (e.g., World War II, Vietnam, Afghanistan)—is that they are mostly set in *invented or fake* scenarios with very limited artificial and stereotypical situations. Games in which the players perform political assassinations (as in *JFK Reloaded* (2004)) or first-degree-murders (e.g., for reasons of greed or jealousy) are nearly nonexistent. The question of why such games are not realized (for lack of player interest or because of moral considerations?) remains open for now.

We were confronted quite unexpectedly with the question of what is morally or ethically tolerable in digital games when we created *PoliShot*, a political Dada game

---

[2]'*Schnelligkeit, Aggressivität, [...] rasche[r] Erfolg und [...] Lebendigkeit*' (Fritz 1995:23).

[3]The catharsis theory (proposed, e.g., by Harvey Carr) maintains that games are played as a means to purge or drain antisocial energy (Carr in McLean and Hurd, 2012:28; cf. Retter 2003:11); it has not been convincingly proven (Kunczik and Zipfel 2010:4).

[4]The interest appears mainly focused on the direction *game to ordinary life*. But Kunczik and Zipfel (2010:10) indicate that people with aggressive personality structures also prefer violent games.

and interactive installation, in 2009. When it was shown in the "Art in Action"[5] and "Computer Art 2.010"[6] exhibitions, it provoked surprising (i.e., shocked and concerned) reactions from curators and visitors alike. The stumbling block was the use of violence, or more specifically, asking the players to act violently in the game. We take this experience as an occasion to enquire into and discuss the contradictions of the actual and the virtual, of concept and content. We attempt to reflect upon the blendings and blurrings of moral/ethical, psychological, and also historical boundaries in digital media, and to (re)trace the influence the computer's particular medial character had and has on these.

2   *PoliShot*: PoliticalGame and ArtMedium

PoliShot was initially created in the university course *Art in Action*.[7] The course addressed practical and theoretical aspects of play, interaction, and art; more specifically, it was focused on digital games, interactive installations, and the Dada art movement.

The participants of the course were asked to develop "Dada games as interactive installations." The games should involve typical Dada ingredients (such as collages, sounds, and sarcasm) and three specific components:

- A mascot, for fun (in *PoliShot*: *Kurt Schwitters*)
- A household appliance as control device (in *PoliShot*: an electric iron)
- A certain *sweetness*, based on individual interpretations (in *PoliShot* the use of cute objects such as a pink swim ring or Mrs. Leyen's braids).

In addition, the games were to be multiplayer games with at least two players, winnable by one of the players (or a team of players), and based on exciting and fast-paced game play. *PoliShot* is designed for four players who support the fight of Dadaists against political lies and political "crimes."

We were offered an exhibition of the student works at the *Weserburg: Museum für moderne Kunst* in Bremen. For this opportunity we revised *PoliShot* from a technical demo, which demonstrated the mechanics of a shooting gallery-style action game: cardboard figures popped up; 1920s gangsters were to be shot; molls

---

[5]*Art in Action* exhibition, Weserburg—Museum für moderne Kunst in Bremen, March 18th–April 5th, 2010.

[6]*Computer Art 2.010* exhibition, with works from the *Goldener Plotter 2010* competition, Innovationszentrum Wiesenbusch in Gladbeck, August 29th–September 26th, 2010 and Städtische Galerie—sohle 1 in Bergkamen, April 1th–July 3th, 2011.

[7]Daniel Cermak-Sassenrath, Bernard Robben, Susanne Grabowski. *Art in Action: Computerspiele, interaktive Kunst und neue Schnittstellen* (Computer Games, Interactive Art and New Interfaces), Course, University of Bremen and University of the Arts (Hochschule für Künste), Bremen, Winter Semester 2009/10.

were to be spared. It was a kind of first-person shooter,[8] involving the basic components of hero/player, opponents, weapons, levels, health, score, and time. This is mentioned because the original game's mechanics provided us with one of the main associations in *PoliShot*: It reminded us of the well-known Dada event *L'Affaire Barrès*. That performance included a stage on which a person symbolized by a puppet was accused and verbally attacked.

2.1   L'Affaire Barrès *and the Context to* PoliShot

In early April 1921, flyers distributed in Paris announced a trial to be held on May 13th. The famous and notorious writer Maurice Barrès was accused of *crimes against the security of the human mind* [*Verbrechens gegen die innere Sicherheit des menschlichen Geistes*] (Hörner and Kiepe 1996:5). The Dadaists Aragon and Breton were disappointed and enraged about the popular writer's exuberant patriotism and the contradictions in his political positions (ibid.:17–24). It was a mock trial, but addressed a serious dilemma: does somebody become guilty who betrays the libertarian ideals of his youth by adopting and advocating conformist ideas only to gain power and influence (ibid.:91). Typical Dada elements in this process are the theme of morals, and the fact that the accused was represented by a puppet; atypical was Dada's role of judge. The trial's accusation was not only directed against Barrès, but also against Dada itself, and the trial became a trial of Dada (ibid.:95).

The ensuing discussions drove Breton to question the future of any revolutionary attitude [*Zukunft jedweder revolutionären Haltung*], a position from which the Dadaist Tzara decidedly distanced himself (ibid.:114). Conflicts between a number of Dadaists escalated and, as a result, several of them turned to surrealism (ibid.:112, 94). The process led to an internal éclat that broke up the most provocative artistic movements of the time.

PoliShot is, in a way, an updated, digital, interactive version of the *L'Affaire Barrès*: we were intrigued by the idea of a public art trial with the accused party being and not being (re)present(ed), and by the possible overlap of art and play. Although not staging a public trial, we developed an interactive installation to be shown in public places. It was concerned with morals, more specifically, political lies; a common everyday topic, we thought of questions of responsibility, corruption, clientele politics, social imbalances, and the like. *PoliShot* was intended as a mock trial against the politicians of our time in which we addressed their lies to protect our human minds. We made public their "crimes" in the areas of social, education, family, environmental, and foreign politics. As in Dada's mock trial, the presence of the accused is not required; indeed, it would get in the way of things. Instead, we developed our own version of proxy puppets. This process was

---

[8]It was a 2D game with a fixed player perspective and unmovable position in the game world; not a 3D world that can be traversed, and so on.

predicated on the notion of transformation across boundaries, that is, mixing references to the historical Dada event with today's politicians, political issues, interests, and positions, and creating a playable game. The game is meant as an "as if," but in contrast to the *L'Affaire Barrès*, participants in the game are asked to act, according to the simple and rigid rules of the first-person shooter: Defend yourself! Shoot and win! The actions of the players are not only supported by visuals, but also by sounds and physical devices.

2.2 PoliShot: *The Dada Game Installation*

by Daniel Cermak-Sassenrath, Susanne Grabowski, and Jörn Ketelsen

The setup of *PoliShot* is shown in Fig. 1. The installation consists of a projection, an ironing board, various input devices, two sets of headphones, and additional items; a Mac *PowerBook* fitted to the underside of the board is running the game software. The interaction devices mix Dada traditions with gaming conventions: players can choose between an iron, a joystick, or a mouse to control the game. The plastic flower, the artificial grass, the swim ring, and the slingshot supplement the setup; these items draw a (nondigital, tangible) connection and create a passage between the game world and everyday life.

Before the game starts, it informs its players about its content, setting, win conditions, and controls (Fig. 2). The navigation is straightforward: mouseclicks or

Fig. 1 The *PoliShot* installation

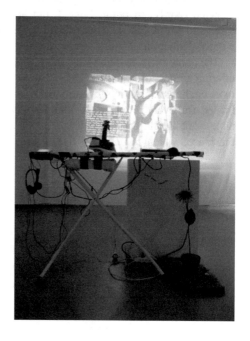

Fig. 2 Title screen

joystick button presses select a scene and a mock weapon. In the game, a left click shoots, a right click reloads. All players do is to select, to (re)load and to shoot, very simple actions that reference, slightly ironically, the conceptual plainness of shooting games (which paradoxically have developed technically way beyond the level of *Wolfenstein 3D*, and have become highly complex and sophisticated).

In the game, Dadaists oppose politicians because they hate liars, depression, oppression, and weapons. They seem to be members of the helpless society for which they standing, but this is just an illusion. The task of the players is to help the artists by silencing the politicians with mock weapons (e.g., a silicone gun) or flatten the cardboard figures with an iron. Attention! There are not only politicians popping up (quickly and easily identifiable by their carrying weapons, and being depicted (partly) in color, Fig. 3), but also Dadaists (unarmed, and displayed in black and white, Fig. 4) whose shooting results in a decrease of truth, articulated as a score deduction.

The game is made interesting and challenging for players of different calibers by offering a range of "weapons" with different properties (Fig. 5). The *slingshot* is the most challenging device for the most daring and skilled artists: it needs to be (re) loaded for every single shot. The *fun gun* shoots six rounds per reload, and is targeted at medium cool Dadaists. The *silicone gun* sprays 40 bullets easily all over the place, which is not very demanding, and every amateur artist can manage.

Fig. 3 Collaged puppets: politicians

Fig. 4 Dadaists

Fig. 5 Selection of mock weapons

Depending on the success of the player, one of two screens is displayed after the game (Fig. 6).

PoliShot draws its subject matter from German politics. We see *PoliShot*'s murderous content as a play on the violent antisocial politics and decisions that were made at the time in Germany. The five political areas addressed in the game are articulated as different levels or scenes (Fig. 7). Prominent protagonists of German politics feature in the game as cardboard puppets, shooting at the player. The puppets of the politicians can be recognized and assigned to political areas through their clothing and props (e.g., the puppet of the minister for family affairs poses in pajamas with a teddy bear). The politicians' collages are fabricated from historical and artistic material; the collages of the Dadaists contain almost only

Hurra, victory!

Defeat

Fig. 6 Win and lose screens

Social politics (Harz IV): Organized poverty for the masses

Environmental politics: More nuclear power to the people

(Non-) education politics

Foreign politics: Oh, nice! Weapons and war

Family politics: Restrict yourself

In-game screen shot (family politics)

Fig. 7 Political areas/scenes

historical photographic material. The figures are made the Dada way as provocative and sarcastic collages. All game objects and scenes are ironically or sarcastically distorted. Even the in-game action with cardboards popping up randomly (Fig. 7, last image) can be seen as complementing the collaged scene. The super mixer computer was used to combine new and old materials, methods, and contents.

PoliShot uses the metaphor that politicians bombard the public with empty speeches and nonsense programs full of lies and contradictions. The players, as members of the public, fight (fire) back. But leaving apart the metaphors, players shoot at human figures that look like well-known politicians. Thus literally taken the game is blatantly violent and its statement dubious.

3 *PoliShot*: Blending Elements and Blurring Boundaries

PoliShot blends Dadaistic, political, and playful elements and blurs their boundaries. Several mixes occur; we observe an interesting moment of interplay between the mixture of formal boundaries and the mixture of content.

3.1 *Blending of* Forms *and* Contents *Blurs* Actual and Virtual *Boundaries*

One of Dada's prominent innovations was the collage. Whereas at first materials such as newspapers, brochures, leaflets, posters, beer mats, and the like were used to create new compositions, soon the photo collage was added to the repertoire. We used the technique of the photo collage extensively to create the visual assets for *PoliShot*. Dadaists enjoyed combining things that ordinarily did not belong together, for example, a woman's head on a man's body. Both parts exist in everyday life, but their combination is a freak. This newly discovered area created entirely new aesthetic possibilities. We located *PoliShot* within this area. The players' recognition of well-known people or objects, such as Frau Merkel's face and the swim ring, invites feelings of familiarity and trust; the unreal composition by collage is irritating and causes feelings of strangeness and distance. Both raise questions of the reality or validity of images, as well as social rules and norms (e.g., showing *Miss* Chancellor in a short dress). They show and open up a possibility of playing with potential but not actual images and actions. It is this play with incompatibilities and contradictions that points us towards new possibilities and suggests ways to overcome restrictive structures.

The computer-supported medium of play even offers participants a simultaneous experience of times, techniques, and worlds that is not available elsewhere. The collage in *PoliShot* emerges as a method of what Bolter and Grusin (2004) term *remediation*. That is, the combination of different times, arts, and media to create a sort of hypermedium that is experienced directly and unmediated by the player (cf. Bolter and Grusin 2004:13). The medium is one of transformative compression.

Another aspect of the abstract, symbolic, or metaphorical representation is the blending of action and content, or a blending of interpretations. We offer players the use of a representation of a mock gun in the game as a means of self-defense, but some people use guns to attack. We use the act of shooting as a metaphor and a functional game mechanic, but some people use the act of shooting as a means to commit crimes. We shoot at collages and caricatures of politicians that are permanent placeholders, but some people shoot politicians.

We were surprised and annoyed by some players' literal reading of the game as a politicians' murder game. But maybe it has to do with a phenomenon Georg

Christoph Tholen (1997) terms a *digital difference*, which refers to the contradictory positions of representation and construction. Media do not simply reproduce copied content, but essentially construct specific aspects of the world (cf. Tholen 1997:115f.). A representation is given a meaning by somebody where, actually, there is nothing to mean, because the image or also the representation of a gun (e.g., a device such as a mouse, joystick, or iron) is far from *being* a gun (see the chapter, "Makin' Cake—Provocation, Self-Confrontation, and the Opacity of Play," in this volume). Nevertheless, an actual situation is constructed immediately, as real as it can be. The medial world of play is considered as a real and unmediated world.

Fritz and Fehr (2003:57) demonstrate that adolescents have their own systems of assessment for physical and virtual violence that are appropriate to their situation of life. They insist on the computer game as a value-free space that adheres to different rules, laws, and principles from ordinary life (ibid). They differentiate clearly between both worlds, much more so than many adults do (ibid.). This position identifies the problem of mixing up actual and virtual worlds as a generational problem. It is probably not the only relevant explanation or possible interpretation, but it aligns with our experiences; the people who most resolutely opposed the showing of *PoliShot* were certainly not young adults.

Games and art can be understood or misunderstood, create or solve conflicts, console or confuse, just as do other media. But what they specifically offer is the opening of associative and interesting spaces for experiences, observations, and conflicts with ourselves and the world around us. These irritations make us become aware of the possibility and necessity of reflecting upon the world and actively changing it at the same time.

3.2 *Blending of* Contents *Blurs* Moral *Boundaries*

At the heart of *PoliShot* are the recognition and the flattening of bogus political programs. For instance, the social reform that became known as *Hartz IV* was described as a program to create wealth and prosperity,[9] but it turned out to be, in fact, the very opposite (which is the topic of the scene in the game, Fig. 7). In the game, the player is asked to interact with the politicians responsible for such nonsense, who continue to offend and "attack" people with their meaningless, misleading, and absurd talk (signified by their carrying different weapons). The game is then understood as a symbolic (gun) battle between participant and politician.

The act of shooting was seen by some players as a dubious, questionable, objectionable, or alarming activity. "I am not going to kill any politicians," as

---

[9]For information on the Hartz IV program which started out as a labor market reform see www.sozialhilfe24.de/hartz-iv-4-alg-ii-2/was-ist-hartz-iv-4.html (in German).

people put it. We were surprised by this feedback, because we did not expect people to focus selectively on some parts of the game while disregarding others, for example, accepting the weapons in the game as guns, but ignoring the ironic collaged images or interpreting the game mechanics as killing, but rejecting the critical artistic/Dada context the game offered. If anybody was metaphorically murdered in the game it was the player and society, and not the other way round! How did such a reversal occur? Or did the fact that it did happen mirror how successful political maneuvers direct people's attention to one aspect while diverting it from another? To focus on the violent side of the game offered an easy way to ignore the rest of it. Or was it the moment of participation that people rejected, being subconsciously aware of their being guilty of active participation or tacit acceptance of making a mess of real life?

We had trouble understanding why some players regarded the digital shooting of cardboard collages with photos of the faces of politicians as a "morally objectionable act." If the installation asked people to "shoot at politicians" and if the work should be removed from the show was debated at length with the jury of the *Goldener Plotter 2010* exhibition. Finally, the jury decided in favor of the freedom of art, and to include the installation in the exhibition.

The controversial discussion prompted us to reflect upon how violence is part of the game. Is it not violent when certain political decisions cause problems and hardship for (some parts of) society? Is violence simply another word for power and potency and are they not everyday aspects or attributes of every society? Are the crummy mock weapons in the game not rather an admission of people's limited individual powers and also an indication of our nonviolent position? Should they not express our powerlessness against power? We believe our game can be seen as an artistic and nonviolent way to express people's dissatisfaction with and their alienation from politics, and to draw attention to its deplorable state. Fritz and Fehr (2003:54f) support our approach when they explain that (actual physical) violence is rooted not in media use but in situations created by society, such as through deceptive political propaganda strategies or cultural repression and suppression. In this case, medial representations follow reality: when people perceive everyday and normal violent reality as unbearable, unacceptable, and morally wrong, commodified medial representations of violence offer a way of compensation, for instance, in computer games (ibid.). Weapons and violence are then the expression of misguided and futile attempts of the players' (self-) empowerment. Violence is not glorified or trivialized but appears as a necessary and appropriate method to gain influence and control in play (ibid:57). *PoliShot* only offers the players the possibility to answer violence with shooting and no other alternatives. It intentionally mirrors the lack of options in ordinary (political) life.

Because of our experiences we asked players specifically about their opinions with regard to moral concerns and discussed the issue of the game's violence with them. We made three observations:

(1) Politicians are granted sovereign rights.[10] When players recognized the faces on the cardboards they came into conflict with a moral code that forbids murder, particularly of members of the government, church, or one's own family (interestingly, nobody either recognized the artists or had quarrels shooting at them). People would feel uneasy if their parents, partners, or children were featured in a violent game. An artistic setting has no relevance in these cases. We have to keep off the political grass, otherwise anarchy looms.

(2) If the figures are not identified or recognized as politicians, for instance, by children, teens, and players not familiar with German politics, people had a great time enjoying the game and no problems whatsoever in playing. For this group of players the virtual representation is object, never subject. Only the game mechanics are relevant: survival outdoes morals (Fritz and Fehr 2003:54). The shift from virtual object to subject is triggered by the recognizable heads on top of the collaged or distorted figures. Some players found the heads problematic, especially when they sympathized with the (real-life) politicians. Fritz and Fehr (ibid.) note that the display of virtual violence can become a problem if it is too closely modeled on the ordinary world.

(3) For most players, the artistic context is not present during the game. Rather players' individual contexts are relevant and employed to assess, judge, and condemn the game. Generally speaking, players with a pedagogical background reacted sceptically or disapprovingly, and players who were professionally or voluntarily involved in politics had bewildered or irritated reactions.

The game is not located in an empty space, and context and frame are not to be disregarded. The game was intended as a work to be exhibited in art museums. It was not designed for children, and it was not distributed for general use. It became obvious that delicate or touchy political or public affairs are observed or examined quite closely and critically, even when they are presented and addressed in an art context, which is generally seen as free and liberal. Why does art appear suspicious when it takes up topics and themes routinely covered by other media? Art was always used as a way to point out and to comment upon problems of society. It would be surprising if this did not include violence.

We were quite unprepared for the intensity and ferocity of the reactions that can be provoked today by an art project. In the following section we therefore deliberate

---

[10]In Philip Roth (1999:24), the protagonist Nathan reports: "And then from talking to me about [boxer] Tony Zale one minute, Iron Rinn was talking to me about Winston Churchill the next .... He talked about Winston Churchill the way he talked about [baseball player] Leo Durocher and [boxer] Marcel Cerdan. He called Churchill a reactionary bastard and a warmonger with no more hesitation than he called Durocher a loudmouth and Cerdan a bum. He talked about Churchill as though Churchill ran the gas station out on Lyons Avenue. It wasn't how we talked about Winston Churchill in my house. ... In his conversation, ... there was no conventional taboos. You could stir together anything and everything: sports, politics, history, literature, reckless opinionating, polemical quotation, idealistic sentiment, moral rectitude... There was something marvelously bracing about it, a different and dangerous world, demanding, straightforward, aggressive, freed from the need to please."

if effects such as these are possibly part of the orchestration of art in our "Society of Spectacle" (*La Société du spectacle*, Guy Debord 1996).

4   Querulousness in Play, Art, and the World Around Us

Is the occasional public excitement or outcry about art merely an act or does it reflect a society's actual moral rules or ethical boundaries? It might prove to be an integral part of the process of how art is produced, perceived, and admitted into popular culture.

Why make art if nobody cares? Why play if things are just as they are in ordinary life? Is art as well as play not predicated on being different from ordinary life? Are freedom and irrationality not paths to places where nobody has been before? Fantasy and insanity drive people to do what can be done: to provoke, to reject, and to show what people could not see and experience otherwise. Art and play are serious, in their own ways, clearly divided from and smack in the middle of everything else, severely limited and dangerously boundless.

Art and play are free from moral obligations and constraints: anything goes! (See the chapter, "Makin' Cake—Provocation, Self-Confrontation, and the Opacity of Play," in this volume.) An artwork or a game can realize things that can or should not be realized in the real world; it experiments without consequences beyond itself. We act as-if, and have a tremendous time even when hundreds of heads roll or cute little lemmings are blown to pieces. We do and we can do because it is possible, and we simply follow what is inside of us, or outside, and it is alright. Or is it not? Although *PoliShot* was finally allowed into the exhibition, its chances of being awarded the jury prize were low, to say the least. This appears to indicate the existence of a blurry line between what is within and what is without accepted boundaries of taste and convention. Traces of this division between good art and bad art can easily be found.

Media artist Jens Stober's first-person shooter *1378 (km)* (Fig. 8) was met with considerable criticism, for example, from the director of the Berlin Wall

Fig. 8 Screenshot *1378 (km)*.
Image: Jens M. Stober

Foundation,[11] Axel Klausmeier. The game's setting is the 1378 km of the former inner-German border. Players can flee the GDR, or ambush refugees as an East German border guard. The game was blamed for featuring crude and degrading content (cf. Berliner Morgenpost 2010). It addresses topics such as the no-man's land, defection to West Germany, and the order to shoot the so-called *Republikflüchtlinge*, which Stober intends to use to generate interest in very recent history. The game's release was planned for the twentieth anniversary of Germany's reunification in October 2010 (cf. Majica 2010), but after much controversial discussion the "serious game" was only released after several months delay (cf. Süß-Demuth 2010).

Similarly to *PoliShot*, *1378 (km)* was quickly accused of being amoral because players engaged in the act of shooting. But much of German history is inhuman and tasteless. Why should this be concealed or hidden by a medium aimed at inviting a critical historical debate? The game is not about slaughtering people and can only be won by not firing a single shot (HfG 2014, cf. Süß-Demuth 2010). The hasty public rejection of the game suggests a political interest in selecting the topics that are suitable for art. For this observation it does not matter if the game was indeed intended as an art work, or merely as a history education project.

It is clearer in the case of Jonathan Meese's work that the reaction to a work of art is an intrinsic part of it. The well-known Meese performed a *Hitlergruß* twice during a panel discussion about art's megalomania [*Größenwahn in der Kunst*] at the University of Kassel just before the launch of the *documenta* 2012 and argued for his signature project, the *Dictatorship of Art* (*Diktatur der Kunst*; cf. hr-online 2013). Predictably, this led to a debate about whether declaring something as art guaranteed a free ride outside the law[12] (cf. ibid.). Following the incident, Meese was actually legally indicted, but later acquitted, because he could convincingly demonstrate that the action was part of a performance and not at all the expression of a political attitude (cf. ibid. and Ackermann 2014). Although the judge indicated that art does not suspend or invalidate the law, she saw Meese's act as a work of art rather than a political demonstration (hr-online 2013). The incident could well have been staged to attract publicity (Reichwein 2013), and political statements appear to work exceedingly well for this. Art is certainly attracted (if not asked) to explore borderline areas (see the chapter, "Playing on the Edge," in this volume). Where one person might use the breaching of morals to invite critical discussion and reflection, another person might mainly or purely seek attention and increased market values (Fig. 9).

There is little doubt about the intentions of Damien Hirst. A trespass of moral values is turned quite directly into monetary valuables. Ulrich characterizes Hirst's work as "not a friendly art which appeal[s] to a majority, but one with which at most only the victorious minority of society can identify" (Ullrich 2011:113, our transl.). Art had become a way of "creating icons of capitalism and celebrating its

---

[11]www.stiftung-berliner-mauer.de/en.

[12]In Germany it is forbidden to show Nazi symbols such as the *Hitlergruß* in public (§86a StGB).

Fig. 9 Jonathan Meese in a typical pose. Image: © Meese, Jonathan: Werke in der Ausstellung. VG Bild-Kunst, Bonn 2016

Fig. 10 Damien Hirst: *For the Love of God* (2007). Image: © Hirst Damien: For the Love of God. Damien Hirst and Science Ltd. All rights reserved/VG Bild-Kunst, Bonn 2016

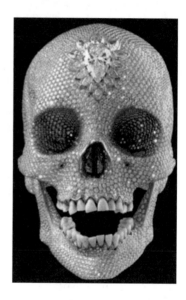

power" (ibid:112, our transl.). Most strikingly, this is celebrated in the work *The Love of God* (Fig. 10). Hirst had a platinum cast of a human skull fitted with 8601 diamonds; the work was produced at a cost of an estimated 50 million British pounds (about 75 million Euros) in 2007 (cf. ibid:91). Any increase in attention that can be directed at such a work can be measured in price: Not *action art* but *auction art*, according to Peter Weibel (2008); "the price tag is the art," as proposed by British journalist Nick Cohen (Riding 2007). But if *The Love of God* was only about (the diamonds') value, why a skull, and why the title? Is it blasphemy? Wolfgang Ullrich (cf. 2011:97) explains that art's *dignity* is fueled by foreignness and divergence from common tastes and norms, and offers a glimpse of a different world. But today the dignity of the transcendence of everyday life has been transformed into a dignity of potency, that is, art not primarily created by artists but demanded by people who have the potential to pay for it[13] (cf. ibid.:98). And this clientele appreciates if the irrational high price of a work is reflected in its motif (ibid.:102f.).

The skull was not selected by accident. As symbol of death, it represents so much existential pathos that can easily be combined with an incomprehensibly high price tag. The project appears to be similar to a potlatch, where the value of a gift indicates the giver's position. Hirst's art is a curious confluence of money, power, life, and death, and a demonstration of his position in the capitalist society (ibid.:102). Finally, Hirst continues a well-established tradition of art: to show people's influence and wealth (ibid:103)—all the artworks described above play in one way or another with structures and balances of power—as *PoliShot* does.

5 Conclusion

Initially, people were drawn to play *PoliShot* by the collaged visuals, and were curious to play it. This was intentional; but starting from the surface of the game, we wanted players to experience and discover on their own the irony, the sarcasm, and also the bitterness of the situations the game is based upon which are to some degree masked. We were attempting this with the means Dada afforded us. Contradictions are used to point out contradictions. Art does not show things that exist anyway but things that are hidden otherwise, says Paul Klee (cf. Klee 1990:76). Art and play refer to the ordinary world and simultaneously distance themselves from it. Both play with the world and against it, and create meaning. Dada's sense was nonsense. This is reflected by the trivial game play of *PoliShot*. It would quite easily be possible to change the game to make it more sophisticated and elaborate, and less offending or more absurd. For instance, the shooting could be replaced by some other action. The sound could be more Dadaistic, cardboards could actually be ironed, bullet holes could be turned into letters and words, and so on. But we like the moment of provocation the original design provides. That is something with which people have to cope. Otherwise there would be a dictatorship of mainstream morality and gentle ideas of decency. Only obvious slapstick would be tolerated to be critical, or established high art, both far removed from everyday experience to avoid treading on anybody's toes or trespassing the boundaries of good taste. Rattattattattattattattatatatattatata! Dadaists aim to bewilder the world, and Dada is the very essence of scandal and provocation.

But the point is not to bemoan on how art is misunderstood in society. It is easy to propagate the independence of art when it is not seen as intended by the artist: context, reactions, and side effects are disregarded. Art becomes actively marginalized. Russian curator Andrej Jerofejew says that Russian politicians would prefer to see modern artists neutralized in some kind of zoo, out of the way, and not

---

[13]Although in this case, it remains questionable if Hirst was able to sell the work successfully.

doing any harm (Rasche 2013). In the times of Dada, art was regularly and severely attacked from all directions with all means and mechanisms. When looking at examples of contemporary art production (Jonathan Meese and Damien Hirst), it appears that today's art critique is in many cases expressed rather politely and moderately if at all. Art is increasingly seen as something disconnected and free from everyday relevance. Does this reflect the tendency to perceive art simply as a thing? But art is more than a thing: it is a system that includes the artist, the process, the work, the reception, and the critique, and is embedded into society. It is part of this system that different parts can contradict each other. This is not a problem to avoid but a challenge to accept, and a game to play.

References

Ackermann, T. (2014). Jonathan Meese durfte Hitlergruß zeigen. In: *Die Welt*, January 2, 2014. Retrieved February 12, 2014, from www.welt.de/123470073.

Berliner Morgenpost (2010). Computerspiel—Schießen auf DDR-Flüchtlinge. Berliner Morgenpost vom, November 29, 2010. Retrieved November 28, 2013, from www.morgenpost.de/politik/article1409769/Computerspiel-Schiessen-auf-DDR-Fluechtlinge.html.

Bolter, J. D., & Grusin, R. (2004). Remediation—Zum Verständnis digitaler Medien durch die Bestimmung ihres Verhältnisses zu älteren Medien. In G. Febel, J.-B. Joly, & G. Schröder, (Eds.), *Kunst und Medialität* (pp. 11–36). Stuttgart: Akademie Schloss Solitude.

Debord, G. (1996). *Die Gesellschaft des Spektakels. Kommentare zur Gesellschaft des Spektakels.* Berlin: Bittermann.

Fritz, J. (Ed.). (1995). *Warum Computerspiele faszinieren. Empirische Annäherungen an Nutzung und Wirkung von Bildschirmspielen.* Weinheim & München: Juventa.

Fritz, J., & Fehr, W. (2003). Virtuelle Gewalt: Modell oder Spiegel? Computerspiele aus der Sicht der Medienwirkungsforschung. In *Computerspiele. Virtuelle Spiel- und Lernwelten* (pp. 49–60). Bonn: Bundeszentrale für politische Bildung.

HfG (2014). Institut für Postdigitale Narrativität—HfG Karlsruhe. *Jens M. Stober (art project)—1378 (km).* Retrieved March 8, 2014, from postdigital.hfg-karlsruhe.de/symposium/jens-stober.

Hörner, U., & Kiepe, W. (Eds.). (1996). *Dada gegen Dada. Die Affaire Barrès.* Edition Nautilus. Lutz Schulenburg Verlag: Hamburg.

hr-online (2013). Hitlergruß als Kunst. Freispruch für Jonathan Meese. hr-online vom August 14, 2013. Retrieved August 21, 2013, from www.hr-online.de/website/rubriken/kultur/index.jsp?rubrik=5986&key=standard_document_49351485.

JIM-Studie (2012). *Jugend, Information, (Multi-)Media: Basisstudie zum Medienumgang 12-bis 19 jähriger in Deutschland.* Medienpädagogischer Forschungsbund Südwest (Ed.). Stuttgart.

Klee, P. (1990). *Das bildnerische Denken.* In J. Spiller, (Ed.) Basel: Schwabe.

Kolokythas, P. (2013). Top 50. Die besten PC-Spiele aller Zeiten. In *PC-Welt online.* Retrieved January 14, 2014, from www.pcwelt.de/ratgeber/Top-50-Die-besten-PC-Spiele-aller-Zeiten-182975.html.

Kunczik, M., & Zipfel, A. (2010). *Medien und Gewalt. Befunde der Forschung 2004–2009. Bericht für das Bundesministerium für Familie, Senioren, Frauen und Jugend.* Bonn.

Majica, M. (2010). Halt—oder wir schießen. Frankfurter Rundschau vom, September 29, 2010. Retrieved November 28, 2013, from www.fr-online.de/kultur/computerspiel-halt—oder-wir-schiessen,1472786,4688516.htm.

McLean, D. D., & Hurd, A. R. (2012). *Kraus' Recreation and Leisure in Modern Society.* London: Jones and Bartlett.

Murray, J. H. (1997). *Hamlet on the Holodeck: The Future of Narrative in Cyberspace*. New York: Free Press, Cambridge: MIT Press.

Rasche, H. (2013). Spaltung in der Kunst—Putins Politik. In: tagesschau.de. Retrieved September 22, 2013, from www.tagesschau.de/ausland/kunstrussland100.html.

Reichwein, M. (2013). Eine Gerichtsposse um seltsame Grußgepflogenheiten. In: *Die Welt*, July 18, 2013. Retrieved February 12, 2014, from www.welt.de/118188302.

Retter, H. (2003). *Einführung in die Pädagogik des Spiels*. Braunschweig: Institut für Allgemeine Pädagogik und Technische Bildung der Technischen Universität Braunschweig, Abteilung Historisch-Systematische Pädagogik, 1998, ext. and rev. ed.

Riding, A. (2007). Alas, Poor Art Market: A Multimillion-Dollar Head Case. In: *New York Times*, June 13, 2007. Retrieved February 13, 2014, from www.nytimes.com/2007/06/13/arts/design/13skul.html.

Roth, P. (1999). *I Married a Communist* (1st int. ed.). New York: Vintage.

Süß-Demuth, C. (2010). Mauerschützenspiel wird doch veröffentlicht. Die Welt vom December 9, 2010. Retrieved November 28, 2013, from www.welt.de/spiele/article11509615/Mauerschuetzenspiel-wird-doch-veroeffentlicht.html.

Tholen, C. (1997). Digitale Differenz. Zur Phantasmatik und Topik des Medialen. In M. Warnke, W. Coy, G. Christoph Tholen (Eds.), *Hyper-Kult. Geschichte, Theorie und Kontext digitaler Medien* (pp. 99–116). Basel, Frankfurt: Stroemfeld.

Ullrich, W. (2011). *An die Kunst glauben*. Berlin: Wagenbach.

Weibel, P. (2008). Der Kult des Auktionskünstlers. In *Die Welt*, October 2, 2008—also cited in Ullrich 2011:96.

Author Biographies

Dr. Susanne Grabowski is based at "DiMeb (Digital Media in Education)" and "compArt: centre of excellence digital art" at the University of Bremen in Germany. She teaches and is doing research in the field of mediating early computer art. For that reason she is working on interactive interfaces, involving artistic, playful, and educational aspects.

Daniel Cermak-Sassenrath is Associate Professor at the IT University of Copenhagen (ITU), and member of the Center for Computer Games Research (game.itu.dk) and the Pervasive Interaction Technology Lab (PitLab, pitlab.itu.dk). Daniel writes, composes, codes, builds, performs and plays. He is interested in artistic, analytic, explorative, critical and subversive approaches to and practices of play. Discourses he is specifically interested in, are play and materiality, play and learning, and critical play. More information is available at www.dace.de.

The Phylogeny of Play

Chris Crawford

Abstract A deeper understanding of the psychology of human play can be obtained by rehearsing the evolution of play. Its roots can be traced back millions of years; with the passage of time, play behaviors became more complex and more closely attuned to specific behavioral needs. Play reached its apex of complexity in Homo sapiens. Understanding this process reveals important lessons about education and game design. Play is a universal human behavior; every culture engages in some form of culturally defined play. The roots of play stretch far back in time.

1 Milestone 1: Locomotor Control

Half a billion years ago, the first vertebrate fish developed a simple means of propulsion: curving their spines alternately left and right, which caused their tails to propel them forward. When the first creatures crawled up onto the land, they used the same basic technique of curving the spine and only now it served to crawl: They curved their spines to the left, placed their right front foot down, and then curved their spines to the right, placing their left front foot forward. In the process, they dragged themselves forward.

Crawling is slow and inefficient; energy is wasted dragging the body along the ground. The next step was to move the legs from the side to underneath the body; land animals graduated from crawling to walking. This saved energy and allowed higher speeds as well.

An animal with four feet can do more than simply walk; it can change the order and speed with which it moves its legs. These different schemes for moving one's legs around are called *gaits*, and there are infinite combinations of the speed and order in which the legs may move. However, there are only a few basic types of gait: In order of speed, these are walking, trotting, cantering, and galloping. This order also reflects diminishing energy efficiency.

C. Crawford (✉)
2349 Sterling Creek Road, Jacksonville OR 97530, USA
e-mail: chrisc@storytron.com

© Springer Nature Singapore Pte Ltd. 2018
D. Cermak-Sassenrath (ed.), *Playful Disruption of Digital Media*,
Gaming Media and Social Effects, https://doi.org/10.1007/978-981-10-1891-6_4

65

The utilization of multiple gaits requires a dramatic increase in the complexity of the nervous system controls of leg motions. Crawling, the simplest of motions, can be programmed directly into DNA; thus, many land vertebrates are able to crawl at birth.

However, the utilization of multiple gaits requires that each muscle control neuron be wired so that, depending upon the gait that the creature is using, it fires at different times relative to other neurons. This requires the addition of extra timing neurons to advance or retard the firing of different neurons depending upon the gait. Such arrangements are necessarily complicated. To handle this, a feature called *spinal pattern generators* has evolved in the spine of all vertebrates (and some invertebrates) that generates the precisely timed signals necessary to control the muscles. It goes without saying that the spinal pattern generators for animals with multiple gaits have more neurons than animals with only a single gait.

Moreover, increasing speeds of movement required faster neurons; it is reasonable to suppose that the speed of neuronal transmission increased over the eons. But eventually, the biochemical limitations of neuronal anatomy imposed a limit on running speeds.

An extreme example of this problem is provided by the cheetah. Running at full speed requires it to place its two forepaws on the ground just 10 ms apart; a mistiming of as little as 1 ms could disrupt the fluidity of the gait and slow or even stumble the cheetah. Thus, the cheetah's overall timing of nerve cells must be accurate to within 1 ms. Yet neurons typically have timing precisions of 11 ms. This limitation is met by grouping large numbers of neurons into sets and using the average timing of the entire set to trigger the muscles.

To understand how that works, imagine a little experiment: Suppose that you want to time with great precision the exact moment when a runner crosses the finish line, and you don't have any electronic equipment, just a lot of stopwatches. You could use one observer who measures with the stopwatch the instant that he sees the runner crossing the finish line, but that observer might be early or late. Suppose instead that you crowded a hundred observers together at the finish line, and each of them pressed their stopwatch button the instant they saw the runner cross the finish line. Some of the observers would be a shade early and some would be a shade late, but if you take the average of all their timings, you'd get a result that is more accurate than that of a single observer. Substitute neurons being fired for people pressing stopwatch buttons, and you get the idea.

Thus, the management of locomotion requires a vast number of neurons wired together in a complex manner that can be altered depending upon gait and speed. Moreover, the behavior of this neural system must alter as the creature grows larger, and later as the creature ages and the muscles weaken. It is possible that such complexity could be encoded into the DNA for a species, but here I propose an alternate hypothesis to explain how creatures managed the complexities of multi-gait locomotion.

I imagine that neurons possessed of three new characteristics developed at some point in evolutionary history. I will not attempt to position these developments with any precision; for purposes of discussion, I will suggest only that they took place sometime after the emergence of terrestrial animals and before the ascendance of mammals, who obviously possessed these characteristics in full form.

The first new characteristic was the ability to alter a neuron's response to inputs based on immediately preceding experiences. It could change the weights that it assigned to different inputs, putting more emphasis on some inputs and ignoring others. In other words, it could change its wiring (from an operational perspective) after the creature was born.

The second characteristic was the development of a new behavior based on random trial and error. Animals could wiggle their limbs around in all sorts of random ways. This new behavior enabled animals to experiment with all sorts of activities, most of which were spastic, but a few of which might prove useful. We see such behavior in all newborn mammals.

The third characteristic was a new system for wiring together neurons into a nervous system. Instead of having every connection planned in advance, the DNA simply programmed neurons to connect to just about everything they could reach. In other words, animals were born (or hatched) with an overconnected nervous system.

We have names for the first two characteristics: learning and play. By playing at moving around, animals were able to teach themselves how to run, walk, gallop, trot, and so on. Learning and play were invented simultaneously, as two aspects of the same development. It is ironic, then, that nowadays we wonder whether play can be used to improve learning when, in fact, play and learning are really two sides of the same evolutionary coin.

This view of the relationship between learning and play is not really so unorthodox: It is just a neurological analog of what genes do in natural selection. Accidental mutations generate random variations in the genes, and those variations are tested against the demands of the environment, rejecting the mutations that do not help, and preferring the mutations that do help. In the neurological analog of this process, play generates random variations in locomotor behavior, which are then tested, not against the environment per se, but against the animal's expectations of efficient and speedy movement. Play is the random generator used to refine the connections in the nervous system. The two primary differences between the two systems are: (1) that the genetic evolutionary process requires many generations to function, while the neuronal evolutionary process—again, what we call learning—takes place over a timescale of days to months; and (2) that the evaluation of effectiveness is external (the environment) in the case of genes, but internal in the case of neurons: There is an internal evaluation process that decides whether the creature is moving well.

This was the first major phylogenetic milestone in human play.

2 Milestone 2: Hunting Play

The second major milestone arose from hunting. Reptiles use ambush hunting, which requires just three steps: (1) wait for prey to come too close; (2) pounce; (3) grapple and kill.

Mammals developed a more complex hunting system comprising five steps. The first is prowling (actively but secretively seeking prey).

The second step is stalking: approaching the prey covertly. This requires a special skill: the ability to put yourself in the place of the prey and calculate what they would see from their position. This permits the hunter to approach more closely without being seen.

The third step is the pounce, which also requires a lot of skill to get the greatest acceleration possible to "get the jump" on the prey.

The fourth step—assuming that the hunter does not get lucky and catch the prey with the pounce—is the chase. Mammalian chases are particularly complex, because the prey knows how to zigzag, which, if the timing is right, will throw the hunter off stride. Of course, the hunter must also anticipate the zigs and zags and compensate for them.

The fifth step is grappling with and killing the prey. This again requires special skills, because most prey has some sort of defensive capabilities: horns that can impale, kicks that can injure, and general thrashing about that, if lucky, can injure an eye. Remember, the eye is the most vulnerable part of the body, and it is only centimeters away from the teeth, which must be brought to bear if the hunter is to kill the prey. How do you get your teeth in for the kill without exposing your eyes to injury? Very carefully, and with lots of preparatory practice: play.

How do carnivores learn these skills? The five basic steps are instinctive, but the details of execution are learned by extensive playing. Carnivores devote most of their early lives to play; it is essential to their survival once they are on their own. This behavior is readily observable in kittens, whose play consists of exercising upon each other the hunting, stalking, pouncing, chasing, and grappling behaviors they will use in adult life. All terrestrial mammalian carnivores have been observed to engage in the same play behaviors as juveniles.

A different kind of play is used by herd herbivores, who need to learn how to evade pursuit by carnivores. Therefore, they play at running, turning, zigzagging, and jumping; we call this play gamboling.

Thus, hunting play was the second major milestone in the phylogeny of play.

3 Milestone 3: Rock-Throwing Play

The third major milestone began more recently, in the last 5 million years or so, when the hominid line separated from the chimpanzee line. The hominids differentiated themselves from the chimpanzees by developing a hunter-gatherer lifestyle.

In this social system, hominids grouped themselves into small roving bands numbering perhaps a few dozen individuals. When a band moved into a new location, it would first set up a base camp. The old adults and the young kids remained at camp, with the old people taking care of the kids and performing domestic duties. The vigorous females would spread out locally, scouring the neighborhood for roots, nuts, berries, fruits—anything edible. They would never stray far from the camp. The males would set out on much longer hunting forays, searching for meat. Initially, they operated primarily as scavengers, grabbing up whatever was left by the primary carnivores. However, competition for carrion was intense, and the males had to fight off other scavengers. The speculation is that somewhere along the way they started throwing rocks at their competitors to chase them off.

It was not long before some genius realized that if rocks could chase off hyenas, they could also kill game. Rocks were not very effective at the task, but hominids combined rock-throwing with two other unique characteristics: bipedal walking, which is particularly energy efficient, and an enhanced cooling system utilizing a superabundance of sweat glands. This allowed them to replace the chase with tracking. Instead of a short, high-speed chase, the hominid hunter tracked the prey each time it outran him, closed, threw rocks that might injure the prey, and forced it to run again. The prey's flight required more energy to traverse a distance than the hominid's steady walk, and the prey's weaker cooling system caused its body temperature to rise. In the hot African savannah where hominids evolved, the prey would eventually be overcome by heat exhaustion and the hunters could slay it.

Rock-throwing provided a valuable augmentation to the process: The hunter who could throw rocks powerfully and accurately could inflict small injuries that, while not fatal, increased the stress on the prey, slowed it down, and shortened the duration of the hunt.

The ability to throw rocks powerfully and accurately conferred a selective advantage, leading to changes in the hominid gene pool favoring that ability. The pectoral muscles gained size and power, as did the biceps. These changes were greater in males, who performed much of the hunting. In modern humans, large pectoral muscles are taken to be an indicator of enhanced masculinity.

Yet musculature alone is inadequate to the goal of throwing rocks powerfully and accurately; the aspiring male hominid also needed to train himself to achieve greater power and accuracy. Thus evolved a male obsession with throwing rocks. Even today, little boys are infamous for their rock-throwing habits. The fanatic devotion to rock-throwing has given birth to a large parade of male sports centered on the hurling of projectiles: baseballs, basketballs, bocce balls, boomerangs, cabers, the discus, footballs, frisbees, hammers, knives, pumpkins, shotput balls, snowballs, spears, tomahawks, volleyballs, and water polo balls.

The pinnacle of rock-throwing, the apex of power and accuracy, is the gun. Nothing hurls a projectile with more power and accuracy than a gun. This explains the fervency with which males in some cultures identify guns with their own masculinity.

This form of play is the third major milestone in the phylogeny of human play.

4   Milestone 4: Manipulative Play

The fourth major milestone in the phylogeny of play began when hominids began using rocks as "unthrown projectiles": By swinging a heavy rock in the hand and smashing it into a target, immense crushing force can be brought to bear on the target. This trick proved to be especially useful for smashing bones so as to extract the protein-rich marrow inside.

At some point millions of years ago, somebody realized that rocks with sharp points were more efficient in this job; this led to selection of rocks for their points. The frequent observation that some rocks cleaved sharper edges upon impact with other rocks provided the insight required for the creation of stone tools, the earliest of which make their appearance in the record almost 3 million years ago.

Particularly striking about these early tools is their fundamental similarity. There is only one basic design: the hand axe, a heavy rock with a single sharp edge, the design of which did not change for ages. Sometime around a million years ago, the record starts to show different types of stone tools emerging, but even then there is still great uniformity of design in any one period and area. It appears that each culture developed its own standard toolkit of stone tools and nobody deviated from that standard toolkit. Apparently people did not play much with stones while making tools.

Over the last million years, the record shows increasing innovation and variety in the hominid toolkits. People were starting to playfully manipulate stones while thinking about their shapes and how those shapes might change under different kinds of blows. People were starting to play with their toolmaking.

This was a huge leap, because it required a major break with the past. All the previous forms of play had been fundamentally motor in their nature. The first form of play, running and jumping to learn different gaits, was confined to whole-body motions and required no real interaction with the environment. The second form of play, mammalian hunting, was also concerned primarily with locomotion, although some interaction with the prey was brought into the mix, introducing a small cognitive element to the play. The third form of play, rock-throwing play, was also fundamentally motor in nature, but it was wrapped up in a complex set of behaviors that required greater cognitive effort. This fourth kind of play, which I call "manipulative play" because it involved use of the hands, was fundamentally cognitive in nature. Yes, it relied upon the hands to manipulate objects being played with, but there was a lot more cognitive activity going on with this kind of play. People were starting to think more about what they were doing, and to think in a playful manner.

5 Milestone 5: Cognitive Play

Sometime around 35,000 years ago, there was a dramatic change in human behavior. Jared Diamond refers to this as "The Great Leap Forward."[1] The archaeological record suddenly bursts with images painted on cave walls, clay figurines, abstract markings in bones and rocks, compound tools, and all manner of other innovations. It is as if humanity suddenly woke up to the creative potential of play. Cave paintings, for example, arose because some curious person wondered what would happen if they daubed mud onto a flat surface in a pattern. Can you imagine the excitement generated when comrades reacted to such stick figures by exclaiming "That looks like a real person!" From there it was a short hop to the cave paintings of Lascaux.

In much the same manner, people began inventing a wide variety of new doo-dads, and in every case we can be sure that creative play lay at the root of many of those discoveries. Did the use of bone needles arise when somebody playfully used a shard of bone to prick something (or somebody)? We cannot know, but we can give short shrift to the hypothesis that bone needles were created by deliberate research.

And what about the use of beads, skin colorings, feathers, and unique clothing to adorn the body? Surely such creations were not conceived from some drive to survive; they were fun to wear, and so began the frenetic race of fashion.

Cognitive play is now the primary manifestation of the drive to play among adults. Adults play with crossword puzzles, game shows, and language play. Detective stories invite the audience to play along by creating miniature stories that explain the crime while fitting the available evidence. As new bits of evidence arise, the hypothetical stories in the minds of audience change. Such stories are really puzzles of story creation.

6 Ontogeny and Phylogeny of Play

Consider this graph presenting an abstract history of the phylogeny of human play: Fig. 1.

While this is obviously a highly simplistic representation, the point of the graph is that, at first, play was exclusively motor in nature, and over the course of time, the cognitive element grew in importance and the motor element diminished. Most of the transition took place very recently, and the motor element has not disappeared.

---

[1]Jared Diamond, The Great Leap Forward, http://wps.pearsoncustom.com/wps/media/objects/6904/7070246/SOC250_Ch01.pdf; recovered February 21, 2014.

Fig. 1 Relationship of motor
to cognitive components of
play with time

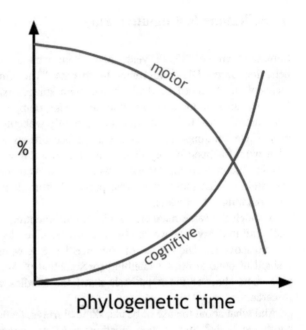

One of the most fascinating ideas to grace intellectual history is *Haekel's Law*: "Ontogeny recapitulates phylogeny." *Ontogeny* is the process of development of a new creature inside an egg or a womb. It is the sequence of developmental stages that the creature undergoes between conception and maturity. *Phylogeny* is the evolutionary path taken by any species. While the overall picture of evolution of life on earth is a bush with ever-spreading branches, the path leading to any single species is linear, and we call that path the species' phylogeny.

There is only one problem with Haekel's Law: It is wrong. Not absolutely, totally wrong, just wrong in too many ways. It is true that there is a rough correspondence between some stages of ontogenic development and some evolutionary stages of that species' phylogeny. For example, every creature starts off as a fertilized egg: a single-celled creature, just as life began with single-celled creatures. During one stage of human ontogeny, the fetus develops gill slits like those in fishes. Later, it develops a tail, which it then loses. So, yes, there are some interesting correspondences, just enough to make Haekel's Law an interesting aside in biology. But there are massive discrepancies as well.

I propose to apply Haekel's Law, in a very broad fashion, to human play. I suggest that the ontogeny of human play faintly echoes the phylogeny of human play. While this suggestion has some obvious problems, there is an undeniable kernel of truth in it. Consider that the human infant's first learning task is crawling—just like *tiktaalik*, all those millions of years ago, crawling out of the sea onto land. Then, the infant learns how to walk, and quickly sets to work playing at other gaits: running, jumping, and so forth.

The second milestone in the phylogeny of human play, hunting play, does not clearly appear in children's play, because we diverged from carnivores a long time ago. Nevertheless, we do see some faint remnants of hunting play. Kids play hide-and-seek games, chase games, and grappling games. Perhaps these forms of play are the last vestiges of our hunting play behaviors.

The third milestone, rock-throwing, occupies a large place in male play, as I mentioned earlier. Because it is a more recent evolutionary development, males never outgrow the impulse to play at throwing things. Note also that this third stage shows up later than the previous two stages.

Lastly comes the fourth milestone: manipulative play. This form of play first appears fairly early in child development, but it does not really kick in with full force until rather later.

Thus, the ontogeny of human play does not follow the phylogeny of play in lockstep; the stages blend together. Yet there remain vague parallels between the two.

7 Lessons

If we apply our ghost of Haekel's Law to play, we can replace phylogeny with ontogeny to produce this graph:

Fig. 2 Relationship of cognitive to motor components in ontogenetic time

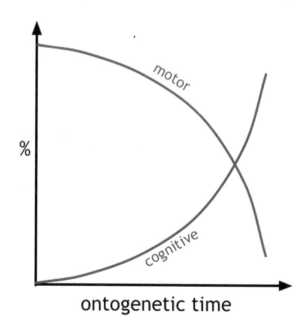

Human play is initially exclusively motor, but as a human child grows, the motor component diminishes and the cognitive component increases.

I earlier observed that learning and play are two sides of the same evolutionary development. If we combine this observation to educational theory in light of Fig. 2, the ideal educational strategy becomes obvious: Education should rely exclusively on motor play during infancy, and then shift toward a greater proportion of cognitive play as the child matures. In fact, this strategy is universal in human cultures. It behooves educators to consider carefully the precise timing of this transition in the educational process.

Game designers can benefit from Fig. 2 as well. We note that play in early childhood is exclusively motor, but acquires a greater proportion of cognitive elements as the child ages. Current games requiring hand–eye coordination retain some of the emphasis on motor elements. It should be obvious that the next step in the evolution of games should entail exclusive reliance on cognitive rather than motor elements. While some games do indeed embody this principle, people have a strong tendency to abandon games as they mature. Figure 2 suggests that game designers could delay that process by placing greater emphasis on the cognitive elements of their designs and de-emphasizing the locomotor elements. In order to tap the adult market, game designers will have to leave locomotor elements out and concentrate exclusively on cognitive elements.

Author Biography

Chris Crawford earned a Master of Science degree in physics from the University of Missouri in 1975. He has published 15 computer games and five books. He created the first periodical on game design, the *Journal of Computer Game Design*, in 1987. He founded and served as chairman of the Computer Game Developers' Conference, now known as the Game Developers' Conference. Chris has given hundreds of lectures at conferences and universities around the world, and published dozens of magazine articles and academic papers. He served as computer system designer and observer for the 1999 and 2002 NASA Leonid MAC airborne missions. His current work concerns interactive storytelling technology. He lives in southern Oregon with his wife, three dogs, seven cats, two ducks, and three burros.

Dingbats Fucktory

Pedro Luis Cembranos

DINGBATS FUCKTORY consists of six videos made up from thousands of *dingbats*, *TrueType* fonts used as ornaments or specific signs in any computer application.

The iconic signs have been transformed and modified to achieve an intentionally incorrect usage which plays off concepts of social relevance and alludes to matters such as economics, politics, everyday life, art and commercial nature. They do so from an ironic point of view, carrying ambiguous messages of the absurd, the paradox, pathos, or plain abjection.

The following 384 dingbats are taken from the posters that accompany the video installation.

P. L. Cembranos (✉)
Madrid, Spain
e-mail: info@pedroluiscembranos.com

DINGBATS FUCKTORY / ECAT - Toledo

Author Biography

Pedro Luis Cembranos Lives and works in Madrid (Spain) and his artistic production studies the strategies that codify individuals within their social sphere and in what measure they configure their most intimate daily acts and nature, by way of narrative systems formed by series or ensembles. His work deals with the questions and social structures that surround the individual and predetermine his or her behavioral patterns, already established by the society and the collective. In a way, his works are usually related to how memory articulates the present, analyzing it as a kind of *historical dream*, making up narrative models in which the familiar-sinister takes the form of parody and pathos. This pathetic attribute of the human being is the one that, to a great extent, codifies man's social acts, building up codes and executing acts that Pedro tries to re-enact and transform in his work.

He likes to work on the marginal, ridiculous, or anecdotal details, and give them the right of the sublime or the relevant, because, more often than not, they are more telling about the nature of man's uncertainties than great evidence or discourse. If he had to, he would define his work as a mental space in which every question is susceptible of forming part of it, one way or another. He is interested in the system's anomalies, the ephemeral quality of everything, the banality of categories, and so on.

Social intercourse is stuffed up with patterns and codes that show, in fact, that reality or the social sphere is yet another form of *construction*. His work plays off the strangeness or even grotesqueness of these patterns that at the same time conform to familiar conventions about normal behavior. His artistic practice frequently consists of works with a tendency to be serious, taking the form of installations, ensembles, or editions, superimposing layers, almost topographically onto the concepts to which he alludes. His work is much related to politics, economics, and culture. He is interested in the *historical document* that every leftover or rejected item or event is, and from a critical position he recurs to ambiguous forms of expression, in which pathos, ridicule, or plain abjection often pop out in his works.

Part II
System, Society, Empowerment

Destabilizing Playgrounds: Cartographical Interfaces, Mutability, Risk and Play

Sybille Lammes

Abstract In this chapter I will examine the triadic relation between play, digital mapping and power. I look at how playing with cartographical interfaces is a central and never neutral activity to digital mapping that invites users to change cartographic landscapes in playful and subversive ways, and thus containing potential to changing the very nature of maps and the spatial relations they invite us to produce. Since the emergence of digital maps, cartography has changed drastically. Digital maps allow for a greater degree of two-way interaction between map and user than analogue maps. Users are not just reading maps but can constantly influence the shape and look of the map itself. Used on our mobile phones, on our computers or as satnavs in our cars, maps have become more personal—transforming while we navigate with and through them. Digital maps have thus altered our conception of maps as 'objectified' representations of space that has been a touchstone for centuries (Anderson 1991; de Certeau 1984; Crampton 2001; Harley et al. 1988). Instead, I will argue in this chapter, maps have become more open to playful, subjective and subversive practices. Play is understood here as a range of activities that go beyond ordinary life by taking on a playful attitude (Cermak-Sassenrath 2013) and as activities of pleasure (Fiske 1993) although not necessarily fun (cf. Malaby 2007). I will probe is where exactly this room to play resides in the particular case of digital mapping and to what extent this gives users agency. Certainly, the image of the map has become mutable and seems to be open to play, but that does not necessarily mean that the power lies solely in the hands of the player/user. How does power work in such ever-transforming neo-cartographies and what affordances does the user/player have to change power-relations?

S. Lammes (✉)
Leiden University, Leiden, Netherlands
e-mail: s.lammes@hum.leidenuniv.nl

© Springer Nature Singapore Pte Ltd. 2018
D. Cermak-Sassenrath (ed.), *Playful Disruption of Digital Media*,
Gaming Media and Social Effects, https://doi.org/10.1007/978-981-10-1891-6_6

1 Digital Mapping

Digital mapping has developed over the last thirty years to become a pervasive and global technology, with powerful relational implications that have reshaped the understanding, production and approach of our spatial world (Thrift 2004). Indeed, a highly urgent question is how particular assemblages of digital mapping change our conception of agency—in other words, our possibilities to develop actions that affect the "outcomes of what the system produces" (Murray 2013). Digital maps—and other forms of data-visualization—allow users to leave traces, tag locations, to find and follow movements, and to trace and connect to others. Above all, digital mapping allows users to see themselves as an intricate part of the map: every move you make is absorbed by the mapping interface. Putting ourselves on the map, and to leave ever-evolving mutable traces of ourselves for others to see, permeates the map with personal visualizations of our movements that can be followed by other people. If we can speak of a shift from "Who am I?" to "Where am I" and "Where am I going" in our public profile, we may critically ask how much control we as users actually have over what traces we leave for others to see and how much say we have over the shape of the map overall.

In this chapter I will discuss how digital mapping interfaces can invite users to put themselves *in* and *on* the map and how this activity can be understood as playful. Play is an important principle in how we use digital maps as a means for socio-spatial networking and how we reconfigure, create and reflect on power relations in spatial terms. I will argue that digital maps should be conceived as specific kinds of navigational interfaces that can proscribe playful performative acts, especially when they entice us to leave traces of our whereabouts on the mapping interface. Being simultaneously signs and things (Latour 1990) they invite users to visually perform and play with their socio-spatial identities that are 'absorbed' by the map as GUI. This ludic quality of digital navigational interfaces needs to be further theorized to understand precisely how digital maps have the potential to proscribe play and how that changes the configuration of our contemporary spatial identities (where am I) in relation to power. So, the fact that digital maps can invite us to inscribe ourselves in the map in a playful manner is key for understanding how much agency players have in constructing such spatial networks. The point that I wish to make here is that digital mapping interfaces allow us to play with spatial identities in the map. This has implication for how we can understand play in relation to power.

The cartographical interfaces that I examine in this chapter are approached as *navigational interfaces*. I use this term to mark a shift in the public perception of maps from the mimetic to the navigational (Lammes 2011). Whilst a mimetic interpretation of maps relies on the belief that maps have a direct resemblance to, for example, a landscape or a battlefield (two points of reference), a navigational understanding approaches maps as outcomes of "chains of production" (Latour 1990) in which references are made depending on relevance. Playful maps underline what November, Camacho-Hübner and Latour have marked as a shift in the public perception of maps since the digital turn (November et al. 2010). They consider risk as a key notion to allow us to move away from an understanding of maps as 'frozen' immutable objects. Yet, I will argue that 'play' is an important and compatible concept to account for this as well and may be even overlapping.

2 The Digital as Ludification

One could actually argue that not only digital maps, but that *all* digital technologies encourage playful attitudes through their interfaces. Computer use is intricately related to play, especially since the 1980 when computer technologies became so much part of our daily lives.

Media theorist Cermak-Sassenrath stresses the capacities of computer technologies in general to engage users in play (Cermak-Sassenrath 2013). As cultural studies scholars have argued before about television and film, media has always enticed users to play as a way for audiences to gain power over the production of meaning (Fiske 1993, 2011; Stacey 1994). Yet digital technologies mark a shift in how we play and engage with power relations through media. Here we can no longer speak of audiences or spectators. Neither can we speak of higher echelons or systems of surveillance that attempt to control media users and to influence their ideological views in covert ways. We have moved from systems that can be described as apparatuses of control (Foucault 1980; Baudry 1976) to what Galloway, following Deleuze, called 'networks of control' (Galloway 2004) that are far more dynamic and distribute power in a more democratic way. The metaphor of representation is no longer sufficient to think about and to understand the relation between power and play in contemporary digital media. Nowadays play refers far more to interactions within fluid networks of media technologies, in which users are embedded as participants. We play with and within these changeable networks.

Such networks are 'navigated' by the user via the interface, a highly important mediator for understanding the relation between power, play and the digital. As Alexander Galloway (REF) points out, interfaces are mediators through which networks come into being and we have to acknowledge their changeability to understand them properly in relation to power. Yet, while Galloway speaks of interfaces as effects, I prefer to speak of them as sign-things in order stress the materiality of interfaces as well as their transformable character (2012). At first glance, this may seem contrary to Galloway's interpretations of interfaces: he speaks of effect as a means to steer clear of an object orientated conceptualisation of the interface which hinders us to think about of interfaces as transformative mediators. Yet the term 'sign-thing' in a Latourian sense doesn't equate with how Galloway comprehends objects (hence the term thing). It goes beyond the object/subject opposition and perceives things as open to change and as having agency: the interface as sign-thing invites users to perform certain actions that are then inscribed in it and become mediated through it. Such a conceptualisation allows us to think of interfaces in terms of changeability whilst at the same time acknowledging their materiality. Moreover, it points to the interface as having agency, an important feature for understanding the workings of power. The question still is, however, what they invite users to do and to what extent this gives users power to play with networks of control and to re-negotiate their shape.

3 Playing the Map: The Mutable Image

In the case of digital maps, this question is foremost related to the extent to which users are able to change assemblages that are mediated through the navigational interface. As navigational interfaces digital maps are mediators that prompt users to form ever shifting assemblages between themselves and other diverse things such as navigation satellite systems, GSM frequencies, base stations, unlocked SIM cards, speed cameras, WIFI signals, computer programs, car mechanics, dashboards, speedometers, roads and other navigators. The mapping interface gives all actants in this network 'signals' to do or perform things: for example, checking our location through a satnav interface prompts the software to seek a GPS signal by connecting to a satellite, which then translates into a refreshed image of the user as part of the Graphical User Interface. This network of humans and things is transformative because the translations between such actants are constantly shifting. The stability of this network is ensured by the immutability of the technologies that together make up this network. These technologies ask us and other actors to perform certain tasks and invite us to act accordingly. "Enter postcode" or "go left" is advice that the satnav offers, which we, in turn, are meant to follow. When we act upon such advise, this is fed 'back' into the network and gets translated into a refreshed image of the map. So our actions change the assemblage that is mediated through the interface.

What is important in terms of power is that the appearance of the interface changes through our spatial interactions with it. The image of the map has become mutable and has become open to play. This mutability seems to be at least discordant with how analogue maps work: here power relations are established through maps as 'frozen' representation that do not change shape easily when being moved around. Particular contexts of use can still make such maps processual in their use (Dodge and Kitchin 2011), but analogue maps as sign-things remain *immutable mobiles*, a term that is highly important for understanding how mapping technologies are produced through networks of asymmetrical power relations.

As a theoretical concept the term immutable mobile was coined by Bruno Latour to understand how power 'works' in producing techno-scientific 'artefacts'. Latour alludes to the story of French explorer La Pérouse to explain what he means by an immutable mobile. In the 18th century La Pérouse was appointed by Louis XVI to travel around the world in order to bring back new information about the explored areas. At one point during his expedition he wanted to establish whether a specific area of China was an island or not, and asked a local inhabitant to draw him a map:

An older man stands up and draws a map of his island on the sand with the scale and the details needed by La Pérouse. Another, who is younger, sees that the rising tide will soon erase the map and picks up one of La Pérouse's notebooks to draw the map again with a pencil... (p.24)

According to Latour there is a crucial difference between the 'project' of the local inhabitants and that of La Pérouse. Arguably the Frenchman has no more knowledge of how to draw a map of this specific area than the old man has, but differently from him La Pérouse wants to be able to bring a map back to France for others to use. The locals have no need for that and can draw maps of their island anytime they want. For them it doesn't matter if maps drawn in the sand are being erased by water or wind. To be able to bring a map back to the king of France, La Pérouse has to make an inscription that is mobile, but also an inscription that keeps its shape when being transported: an immutable mobile. An immutable mobile is a *flat inscription* that can vary in *scale*, can be *reproduced*, is *re-combinable* and is *super-imposable* with other inscriptions (37–38). When maps become immutable and mobile, they acquire a certain authority and it becomes more difficult for users to undo or change them. In the case of La Pérouse the map becomes an immutable mobile so the King of France can use it as a powerful representation in his quest for world domination.

One could argue that maps have become even more mobile in the digital age. They emerge in "flux" (Hayles 2002) with people becoming increasingly hyper-mobile. Yet through this spatio-temporal acceleration also acquired a certain degree of adaptability or mutability. This mutability manifest itself most clearly in the image of the digital map, which changes its visual appearance according to where we go and what we want to see. Now the map-user has a certain say in how scales vary (zooming in) and which images are combined and superimposed (layers, mash-ups): we can play with the image of the map that has become mutable.

In spite of this, digital maps still depend on the practice of inscription. This is most notably the case with Google Earth. It is actually a 3D digital globe on which a multitude of inscriptions are superimposed. Perfectly in line with Labour's def-inition, the globe itself and its basic cartographical features are immutable, yet super-imposable and re-combinable. The views and degree of zooming in and out has spectacularly increased in the case of Google Earth, but as a tool and toy it actually still heavily depends on reproducible inscriptions. It is in that sense—in concordance with Latour's claim (1997)—that the term immutable mobile has not been made redundant since the digital turn, although velocity may have been increased tremendously and other connections may have been privileged:

(I)n the long history of immutable mobiles, the byte conversion is adding a little speed, which favours certain connections more than others, than this seems a reasonable statement. To say that we are living in a cyberworld, on the other hand, is a complete absurdity. (n.p)

Indeed, one could state that in Google Earth the practice of hybridization, which has always existed according to Latour, is sped up and augmented to a far greater extent and also made more apparent than in the case of analogue maps. New kinds of connections can be established (e.g. webcams, photographs) and the rate at which images can be added and re-combined has accelerated. Yet, in essence, the images that are re-combined via the interface are still re-producible inscriptions and thus curtail the possibilities to renegotiate asymmetries.

An open source mapping application like OpenStreetMap (OSM) also depends on a multitude of visible and re-combinable inscriptions. Users can zoom in and out and can enrich the map with existing layers for walking, cycling or driving. Like in Google Earth or Google Maps the image of the map is also arranged according to certain pre-determined gridlines that cannot be changed. Yet, in OSM, the mapping interface is definitely more mutable than in Google Earth because the user is now actively encouraged to contribute in-depth inscriptions to the map. Or as the opening webpage puts it: "OpenStreetMap is a free worldwide map, created by people like you." This suggests an input of users that goes much further than the activity of layering, such as being used in the Google Earth project "Save the Elephants" in which the mobile GPS traces of Elephants are overlaid on the Google Earth globe surface, to be removed by users at their wish. OSM users are invited to add immutable map inscriptions instead of only adding layers. They can make updates that change how the map looks as an inscription and have therefor more power in how 'the world' is viewed. Another good example of this is WikiProject Gaza where OSM mappers changed the map of the Gaza strip to improve humanitarian relief (OpenStreetMapWiki). Users thus have possibilities to become explorers and cartographers who can alter the map by inscribing changes. The traces they leave cannot be easily removed. This position of OSM mapper actually somewhat evokes that of the young Chinese men in Latour's story that makes a drawing of the island in La Pérouse's notebook for him to take back to France. Similar to this young man, OSM users that are not necessarily map experts are encouraged to make map inscriptions and to become mediators or translators. Since contributors of to the map make these alterations in the surface instead of on it, cartographical images become less asymmetrical inscriptions and regain at least a taste of mutability.

4   50 Shades of Play

Play is an important feature in how users can engage with OSM as mappers, both in the sense that they are asked to make use of the play in the map (its mutability), as in how such inscriptive endeavours are shown to others. In addition to having possibilities to play games to help with developing map inscriptions (e.g. Address Hunter), OSM mappers also engage in "performative play" (Sutton-Smith 2001) through their direct cartographical engagement with the mapping project (an activity that is compatible with what Nitsche (in this volume) describes as crafting). Furthermore, diaries, blogs and efforts to help the OSM community are rewarded with badges and scores. Mappers can earn bonuses for "auto-biography", "citizen patrol", "clean up" and "editor". So play is an important activity in how mappers show themselves to the OSM community as cartographers, explorers, travellers, climbers, walkers, runners and artist. Most importantly, though, OSM mappers leave traces of what they have changed *in* the map for others to see. This is done through the option of "GPS traces" and by looking up the name of a contributor to see what she exactly did for the map and which inscriptions were left when and where. Unlike a conventional analogue map where such inscriptions are 'depersonalised' and we cannot easily know which assemblages of actants established it as a thing, upward chains of production are partly traceable and even celebrated in OSM.

When it comes to understanding the triad relation between power, play and the mapping interface, the potential of leaving traces in the digital map is crucial. OSM makes these traces part of the inscription of the map, but most digital maps invite us to put play in the map as well as to putting ourselves as players in the map (Lammes 2013). Mapping devices may come in many shapes and forms and their functions may be highly diverse, yet one thing most of these maps share, and which makes them ontologically and epistemological profoundly different from analogue maps, it is that the user can put herself *in* the map for others to be seen.

In particular maps that are used in mobile settings and/or are part of a social networks such as Facebook or Foursquare, persuade users to put themselves in the map and play with their spatio-public image. This changes how we are in, how we shape and how we know the 'world.' So instead of looking at a map and maybe even putting markers on it to *represent* your movements, the map now *simulates* your movements in a procession manner. It does no longer, as Gekker and Hind puts it, "relegate the map to a secondary level underneath the real (…) world" (Gekker and Hind 2013). You have become part of the map, and the map constantly 'absorbs' your material whereabouts (Lammes 2011). Surely a digital map used when driving has a different and 'lower' playful function than a mini-map in a computer game or mapping as part of a locative artwork. Yet all digital maps, so also satnavs, invite us to play to a certain degree, be it in a more subtle or overt ways. Satellite navigation interfaces may at first seem to be rather remote from play in their purpose, but in more subtle ways play is part of our navigational experience through celeb voices (Patsy or John Cleese, for instance), racing flags to indicate that you have reached your destination, and through the sheer similarity between the look of the screen of your satellite navigation devices and a game such as Grand Theft Auto (Chesher 2012). But play is foremost present in how we interact with the navigational interface: the conversation we can have with the satnav ("No, Katie you're wrong, we have to go right here"), the way we can be amused to see ourselves end up in unexpected places on the screen, and, last but not least, how we can race against the satnav while looking at ourselves on the screen. According to a UK survey, 7 million car drivers tried to beat the time of the arrival estimated by their satnavs, a rather dangerous game that shows how closely related risk and play are, and how we love to play the system.

5   Putting Players in the Map: Risk, Power, and Play

In their article "Entering a risky territory: Space in the age of digital navigation" November et al. (REF) assert that digital maps accentuate that risk is *in* the map. It can only be conceived as part of the map when we conceive maps as navigational instead of mimetic 'mirrors' of reality. In the case of digital maps users are invited to be navigators and are encouraged to approach maps in terms of the risk assessment. The example of racing against the satnav illustrates this perfectly: we interact with the interface, and by taking the risk to go against its advice we become aware of how this technological assemblage makes certain references prevail over others. Furthermore, an obstacle on the map, such as a traffic jam, is estimated in terms of risk. The same holds when we end up in the wrong place, which points to the risk of being too late, or more specifically, the incalculable and unpredictable outcomes of this chain of production that is translated via the navigational interface (or what the authors call a "dashboard"). This makes the map user aware, as November et al. argue, that (digital) maps don't depend on one singular indexical relationship. It also makes it possible for users to get some understanding of how chains of production are set in motion, hence revealing how they are networks of connectivity, rather than fixed structures. Yet, when we are using technologies in mundane settings, risk seems to be a rather heavy term. Yes, we may take the risk of getting a fine when speeding or running through a red light, but the physical risk is most often limited, especially when safety rules aren't being broken. Play may be a complementary term to risk, since it also points to the fact that digital maps are outcomes of processes of translation that are by no means mimetic and pre-calculable, but it also includes interactions with digital mapping interfaces that are not necessarily dangerous or 'deep play' (Geertz 1972). It acknowledges that digital mapping interfaces invites us to put play in the map, both in more or less dangerous settings. Furthermore, it also acknowledges the ritualistic side of this navigational mode of being.

6   Deep Play, Open Play and the Power of Tinkering

Maybe we can conclude that deep play correlates with a higher degree of power of the user over the shape of the network than more safe kinds of play. When an interface actively invites users to intervene in the system, such as OSM does, it makes play more dangerous. To become an active mapper, as the Gaza contributors did, you may have to go exploring more hostile or remote areas through walking, climbing, or sailing than when engaged in more safe ways of playful navigation. With the mapping device in hand as play equipment, mappers put a higher degree of risk in the map for others to see. The news item of the satnav racers makes users aware that they are part of this fluid network, but also of where their agency in shaping this network stops. The more play leans towards deep play, and puts risk in the map, the more possibilities it gives us to play as a means for appropriating and shaping power.

However, the main rules of OSM are more difficult to change. The base map still functions according a dominant Western cartography, a "Cartesian-Newtonian epistemology informed/transformed within both historical and current (…) colonial projects of the West" (Johnson et al. 2006). What is up and down, what are borders, what are distances: such structures remain more difficult to challenge, even in OSM. Thus, the degree of play is reduced by the fixed basic structure of the map that is very much ideologically informed and shapes our understanding of spatio-temporal relations.

The wide range of contemporary mapping interfaces may entice us to put play in the map in more-or-less perilous ways, but they all lack the openness to change the basic map itself. Contemporary digital maps do not entice users to engage in activities that combine deep play with *open* play. Although digital maps may hybridize mapping and touring (de Certeau 1984) the navigational interface leans heavily on an ideologically informed 'rational' base-map that limits the agency of users in making mapping a practice of their own, in tune with how they may want to produce and understand spatial relations. A navigational interface that enables users to play the system to its full potential, should both invite users to engage in deep play by making inscriptions in the map, as well as to adapt the relational structures on which that map is based. This may be in the shape of digital counter-mapping or "vernacular mapping" projects (Gerlach 2010), which go further than some current participatory mapping projects based on a fixed map structure which is then layered or sometimes inscribed with geo-narratives (Pyne and Taylor 2012). But one could also envisage a game with a different kind of mapping interface—one that would stimulate another kind of involvement that is, for example, far more in tune with how people cognitively draw maps in their heads while moving or how indigenous people dream landscapes (Hirt 2012). Such digital interfaces would go one step further in encouraging users to take agency in thinking and producing their 'Umwelt'. (Thrift 2005; see also Khaled in this volume)

As I have shown in this chapter, digital mapping technologies open up several possibilities for playing with spatial relations. Computer technologies made maps into interactive interfaces that are far more susceptible to play than analogue maps are. Although critical geographer Chris Perkins (2009) rightly argues that mapping interfaces have always been open to playful conduct (of which many board games testify) digital mapping interfaces are significantly more playful than analogue maps because of the high degree of transformative spatial interaction between actants that they mediate (Perkins 2009). This spatial interaction is even more pronounced when it results in moving and mutable simulations of the users in the map. Navigational interfaces that invite users to go beyond layering the mapping image and in addition encourage them to engage in deep play, have the potential to subvert the networks of control that are mediated through the navigational interface. Yet digital maps should even become more open to counter-play to really take control of such networks.

Acknowledgements The research leading to these results has received funding from the European Research Council under the European Community's Seventh Framework Programme (FP7/2007–2013)/ERC Grant agreement no 283464.

References

Anderson, B. R. (1991). Census, map, museum. In *Imagined Communities: Reflections on the Origin and Spread of Nationalism* (pp. 163–185). London: Verso.

Baudry, J. L. (1976). The apparatus. *Camera Obscura 1*(1).

Cermak-Sassenrath, D. (2013). Playful computer interaction. In V. Frissen, S. Lammes, M. Lange, J. de Mul, & J. Raessens (Eds.), *Homo Ludens 2.0: Play, Media, Identity*. Amsterdam: AUP.

Chesher, C. (2012). Navigating sociotechnical spaces: Comparing computer games and satnavs as digital spatial media. *Convergence: The International Journal of Research into New Media Technologies, 18*(3), 315–330.

Crampton, J. W. (2001). Maps as social constructions: Power, communication and visualization. *Progress in Human Geography, 25*(2), 235–252.

de Certeau, M. (1984). *The Practice of Everyday Life*. Berkeley, London: University of California Press.

Dodge, M., & Kitchin, R. (2011). Mapping experience: crowdsourced cartography. Available at SSRN 1921340.

Fiske, J. (1993). *Power Plays, Power Works*. London, New York: Verso.

Fiske, J. (2011). *Reading the Popular* (2nd ed.). London; New York: Routledge.

Foucault, M. (1980). *Power/Knowledge: Selected Interviews and Other Writings, 1972–1977*. New York, N.Y.: Pantheon Books.

Galloway, A. R. (2004). *Protocol: How Control Exists After Decentralization*. The MIT press.

Geertz, C. (1972). Deep play: Notes on the Balinese cockfight. *Daedalus, 101*(1), 1–37.

Gekker, A., & Hind, S. (2013). Fingertips and foot pedals: The casual nature of digital mapping. In *From Pole to Pole, ICC 2013, 26th International Cartographic Conference*. Dresden. http://icaci.org/files/documents/ICC_proceedings/ICC2013/_extendedAbstract/137_proceeding.pdf.

Gerlach, J. (2010). Vernacular mapping, and the ethics of what comes next. *Cartographica: The International Journal for Geographic Information and Geovisualization, 45*(3), 165–168.

Harley, J. B., Cosgrove, D., & Daniels, S. (1988). Maps, knowledge, and power. In *The Iconography of Landscape: Essays on the Symbolic Representation, Design and Use of Past Environments* (pp. 277–312). Cambridge Studies in Historical Geography 9. Cambridge: Cambridge University Press.

Hayles, K. N. (2002). Flesh and metal: reconfiguring the mindbody in virtual environments. *Configurations, 10*(2), 297–320.

Hirt, I. (2012). Mapping dreams/dreaming maps: Bridging indigenous and western geographical knowledge. *Cartographica: The International Journal for Geographic Information and Geovisualization, 47*(2), 105–120.

Johnson, J. T., Louis, R. P., & Pramono, A. H. (2006). Facing the future: Encouraging critical cartographic literacies in indigenous communities. *ACME: An International E-Journal for Critical Geographies, 4*(1), 80–98.

Lammes, S. (2011). The map as playground: Location-based games as cartographical practices. In *Fourth International Conference of DIGRA*. http://www.digra.org/wp-content/uploads/digital-library/11310.35282.pdf.

Lammes, S. (2013). Digital cartographies as ludic practices. In J. Thissen & K. Zijlman (Eds.), *Understanding Contemporary Culture: New Directions in Arts and Humanities Research* (pp. 93–100). Amsterdam: AUP.

Latour, B. (1990). Visualisation and cognition: Drawing things together. In M. Lynch & S. Woolgar (Eds.), *Representation in Scientific Activity* (pp. 19–68). Cambridge: Mass.

Murray, J. (2013). Agency|Janet H. Murray's Blog on Inventing the Medium. Retrieved May 2013 from http://inventingthemedium.com/tag/agency/.

November, V., Camacho-Hübner, E., & Latour, B. (2010). Entering a risky territory: Space in the age of digital navigation. *Environment and Planning D: Society and Space, 28,* 581–599.

Perkins, C. (2009). Playing with maps. In M. Dodge, R. Kitchin, & C. Perkins (Eds.), *Rethinking Maps* (pp. 167–188). Taylor & Francis.

Pyne, S., & Taylor, D. R. F. (2012). Mapping indigenous perspectives in the making of the cybercartographic Atlas of the lake Huron treaty relationship process: A performative approach in a reconciliation context 1. *Cartographica: The International Journal for Geographic Information and Geovisualization, 47*(2), 92–104.

Stacey, J. (1994). *Star-gazing: Hollywood Cinema and Female Spectatorship.* London, New York: Routledge.

Sutton-Smith, B. (2001). *The Ambiguity of Play.* Cambridge Mass: Harvard University Press.

Thrift, N. (2004). Movement-space: The changing domain of thinking resulting from the development of new kinds of spatial awareness. *Economy and Society, 33*(4), 582–604.

Thrift, N. (2005). From born to made: Technology, biology and space. *Transactions the Institute of British Geographers, 30,* 463–476.

Author Biography

Sybille Lammes is professor New Media and Digital Culture at the Centre for the Arts in Society (LUCAS), Leiden University. She has been a visiting Senior Research Fellow at The University of Manchester, and has worked as a researcher at the Centre for Interdisciplinary Methodologies at the University of Warwick, as well as the media-studies departments of Utrecht University and the University of Amsterdam. Her background is in media-studies and game-studies, which she has always approached from an interdisciplinary angle, including cultural studies, science and technology studies, postcolonial studies, and critical geography. She is co-editor of Playful Identities (2015), Mapping Time (2017 fc.) and The Routledge Handbook of Interdisciplinary Research Methods (2018 fc.). and The Playful Citizen (2017 fc.). She is an ERC laureate and has been the PI of numerous research projects. She is a member of the Playful Mapping Collective.

Crafting Through Playing

Michael Nitsche

Abstract Through productive play, the process of playing itself is reframed as a form of crafting. The essay explores the context for playing as crafting as it draws from craft research and game studies to present a different view of emergent play. Huizinga's definition of play serves as a starting point into a shift to craft-like practices, which are illustrated with a discussion of selected machinima work that serves as example for this concept of playing as crafting. Finally, the overlap between playing and crafting is discussed as an example for critical making.

Keywords Craft · Emergent play · Machinima · Critical making

1 Introduction

Nature scientists do not take authorship lightly, particularly not in publications with high impact factor. But there they were: Foldit Contenders Group and Foldit Void Crushers Group, two groups of what might be best described as gamers' guilds had become co-authors of a paper in Nature Structural and Molecular Biology (Khatib et al. 2011). Both were included in the announcement of the protein-folding discovery made with the help of *Foldit*, a game custom-built to address this particular scientific problem. Even though these players "have little or no background in biochemistry" (ibid.: 1175), they still crucially contributed to relevant discoveries regarding certain protein structures. The achievement of the game *Foldit* was to make the research problem accessible to players without the need to fully educate them about the necessary background or scientific complications. Players created a solution through the scaffolding provided by the game, not through learning the underlying scientific conditions. Instead, these conditions were implemented in the game's functionality.

M. Nitsche (✉)
Georgia Institute of Technology/Digital Media, 85 Fifth Str. NW TSRB,
Atlanta, GA 30332-0165, USA
e-mail: michael.nitsche@gatech.edu

© Springer Nature Singapore Pte Ltd. 2018   99
D. Cermak-Sassenrath (ed.), *Playful Disruption of Digital Media*,
Gaming Media and Social Effects, https://doi.org/10.1007/978-981-10-1891-6_7

Success stories like these show the importance of play as productive activity, and they motivate this essay, but they are not at its core. In the case of *Foldit*, the question was not whether the protein folding happened—scientists were aware that the structure had to perform the fold—but how it happened. In this way, *Foldit* was a puzzle game for which the beginning and end conditions were given but not the solving process in-between. No doubt, new knowledge was generated and the discovery of processes between known conditions is a key component of science. It is not key to understanding production under the auspices of craft, though. To create its "output" craft applies knowledge through skillful and creative practice. It might be highly repetitive at times, but it is always about creating something new, not solving an existent puzzle. This chapter sets out to investigate this creative production as opposed to pre-structured productivity. To highlight the differences and overlaps, it will move along the dividing line between these two forms of productive play to discuss the difference.

2 Mapping the Field

Any discussion of play as craft stands next to ongoing debates on the relationship between play and the "real world". Scholars have long investigated the effects of game actions and their reach beyond virtual domains. Their value for social bonding has been widely discussed (e.g., Taylor 2006); complex blends between social and economic interests have been traced (e.g., in the principle of "guanxi" between Chinese *World of Warcraft* players Lindtner 2009); focusing completely on the material side, others have explored ecologies in game worlds (Castronova 2005) and emerging legal complexities in them (Lastowka 2009). Social, financial, and legal issues have transcended from physical realms into virtual ones and back to encompass both into a wider world of mixed materiality. As long as you can make a living from selling virtual clothes in *Second Life*, productive tools can be software as well as needles and sewing machines.

Using these tools leads to another reference point for craft in games: the transition from player to modder to coder and the countless states in-between (Camper 2005). Particularly, Pearce (2006) emphasizes the craft-like production through emergent play. She bases her argument on three tiers: (1) creative empowerment of players "has become a viable business model" (2) "the malleability, discursive quality, and networked infrastructure of the Internet return us to a pre-industrial culture of play" (3) productive play is connected to a postmodern transmedia experience "in which appropriation is not only allowed, it is exalted" (Pearce 2006). She contents that modern media conditions emphasize productive play, which points back to a kind of revivalist pre-industrial ideal, but still pays. This is supported with examples from the modder scene. A community of modders can contribute substantial value to an existing game (as seen in the *Counter-Strike* mod for Valve's *Half-Life*) or even facilitate a game's sheer existence (as seen in player-run servers of the otherwise canceled *Uru* universe discussed by Pearce).

Here, the practice of play as craft is exemplified by the content creation and modding behavior of player communities. Many of their means of digital production, such as modeling a new 3D game level or coding additional elements, have already been associated with craft activities. Digital production methods through advanced interface options mixed with other creative options have led to an emerging area of "digital craft".

"Think of the digital artifact, shaped by software operations, made up of data assemblies. [...] It is abstract: a symbolic structure, a workable construction, in a digital medium, showing the effects of manipulation by software tools" (McCullough 1998). Craft itself is being redefined and applied to the digital production.

> [C]raftspeople can be defined generally as people engaged in a practical activity where they are seen to be in control of their work. They are in control by virtue of possessing personal know-how that allows them to be masters or mistresses of the available technology, irrespective of whether it is a mould, a hand tool, an electrically driven machine or a computer. It is not craft as 'handcraft' that defines contemporary craftsmanship: it is craft as knowledge that empowers a maker to take charge of technology (Dormer 1997).

On the flip side, game techniques have been applied to engage people not necessarily in the game as an ends of all means but to motivate them in other activities. As *Foldit* shows, gamers do not need the full knowledge of the underlying system but the game-like design can channel the play into productive channels. The trends of "serious games" and "gamification" heavily loom over the topic.

The various fringes of productive play ask for an outline what this text is about and what not. They also already foreshadow that the text will weave back in and out of these territories to develop its argument. To explain the idea of crafting through playing, one has to clarify our understanding of what defines a crafting activity.

First, this text will be about forms of play that lead to some object as outcome. Pure gain of knowledge—as achieved in a successful educational game—is not included as a quantifiable object-outcome. Gain of knowledge through practice is indeed an effect of craft, a practice that has long embraced a "learning by doing" tradition, but new skill is gained through production not as the single targeted outcome of it. Furthermore, the resulting object has to be independent from the game system and self-contained. That means that the productive outcome's value is not diminished when the game world is switched off; much like a piece of pottery does not lose its value when it leaves the workshop. For example, gold farming in *World of Warcraft* depends on the operation of the game world and does not qualify, neither do titles such as *Little Big Planet*, *SimCity*, or *MineCraft* that offer laudable creative options but these options remain active only within the game system.

Furthermore, I will attempt to only talk about instances where this form of productive play happens for the benefit of the individual player. That means that tricking the player into a play behavior to provide workforce for a third party will be seen as a particular case but not as the core for this exploration at hand. For example, if players do not understand their activity in *Foldit*, then it shall remain outside our discussion here, just like many forms of "gamification" that hide productive activity under a cloak of game mechanics. A system such as the *ESP game* that uses players to tag meta-data to images "without realizing they were doing so" (Ahn and Dabbish 2004) is completely excluded.

The goal is to investigate play that leads to the production of a self-standing quantifiable outcome, an object in its own right and by the actions of the player that emerges as independent from the game system. This particular investigation is not meant as criticism of other approaches—it is very laudable to generate new knowledge through play, for example—but is meant as an exploration of the fringes of play as creative production. It turns to games as tools and play as craft in the hope to find a new critical perspective on the borderlines of play.

This conflicts with some definitions of play that concentrate on fragments of Huizinga's definition of play as "an activity connected with no material interest and no profit can be gained by it" (Huizinga 1949). A discussion of the here-applied interpretation of play will be provided first to counter such critique. From there, the argument will draw from craft research as well as game studies to build bridges between the two. Craft research focuses on craft's social, historical, and theoretical background and provides a critical access point to craft practices. On the other hand, game studies have readily embraced the notions of art and particular of design, but its relationship to the third sibling of creative practices, craft, remains less explored. Given the growing popularity of Maker and DIY cultures, the underlying notion of craft sees somewhat of a resurgence.

New technologies that blend the digital with the physical build hybrid bridges that allow us to cross over and apply the debates from traditional and not-so-traditional craftwork to video games.

Secondly, the text provides an example where video game play is being transformed into productive activity. The practice of machinima production will illustrate how this activity materializes and to what results it might lead.

Finally, the argument turns to critical making as a different lens to the concept of critical play. The goal is to provide a different approach to designing play (and consequently games), one that incorporates a productive quality in a possible transformation of play to craft as a form or self-expression and of games to tools for craft-like production.

3 Playing as Crafting

Video games and game studies have largely concentrated on their connection to design and art. "Game design" has become a professional job description and an academic field of study; the discussion of "games as art" has incited different communities for better or worse and spawned a lively debate as well as own publications and events. In commercial game production, the credits of most titles distinguish between the designer for the game and the artists creating the assets needed for this title. While the game designer faces many tasks typically associated with interaction design, the task of the game artist is far closer to that of a traditional artist or craftsperson. It is only through a successful collaboration between these different parts that players of the resulting video game will experience its virtual environment as meaningful, functional, expressive, and carefully assembled. The principal design (the particular kind of game play it applies), the artwork (the aesthetics of the game), and the crafted details (including the environments and other assets) have to work together to form a successful and engaging result. There is no successful video game production without a good portion of craft.

Craft finds itself repeatedly enthralled in a debate to defend its role in the face of its more visible siblings (Shiner 2010) and tracing its own roots (Adamson 2013). One side effect of these debates is that craft research gains traction in digital culture as an increasing number of scholars trace connections between digital media and craft. They investigate, for example, the question of production and materiality in a culture that nowadays includes 3D printers and other digital facilities (Bunnell 2004), how it changes education (Bonanni et al. 2008), or production (Gershenfeld 2012). If individualized production is part of the next digital revolution, then craft is needed to understand it.

In return, the connections are also evolving from traditional craft toward digital media: Novel forms of physical computing have inspired new craft techniques, and some forms of creative coding and interactions are already seen as a form of craft that lives exclusively in the digital domain. For example, mastery of coding or 3D modeling has been noted as a form of crafting within the craft literature (McCullough 1998; Masterton 2007).

Building on these connections, we follow this turn to the individual player-producers and turn to the practice of play as craft as the critical perspective. Craft is used mainly to look at game players, not game producers. But, as will become clear, the edges will once again blur depending on our understanding of "play".

Huizinga's work on play has been used, debated, and reinterpreted countless times over, but the role of games that facilitate play as cultural practices remains largely unquestioned. As Salen and Zimmerman put it:

> Stated simply, games are culture. (…) games offer players forms of participation that extent the boundaries of play beyond the edges of the magic circle. From player-produced objects like skins, mods, or game patches, to role-playing games in which players explore and alter their personal identity, games have the potential to transform culture. These cultural transformations emerge from the game to take on a life of their own outside the framework of game play (Salen and Zimmerman 2004).

The importance of play as form of cultural production is already inherent. The much-cited summary of Huizinga's definition of play is at the root of Salen/Zimmerman's view. It reads:

> Summing up the formal characteristics of play we might call it a free activity standing quite consciously outside 'ordinary' life as being 'not serious', but at the same time absorbing the player intensely and utterly. It is an activity connected with no material interest, and no profit can be gained by it. It proceeds within its own proper boundaries of time and space

according to fixed rules and in an orderly manner. It promotes the formation of social groupings which tend to surround themselves with secrecy and to stress their difference from the common world by disguise or other means (Huizinga 1949).

Notably, this does not exclude production per se but instead rejects "material interest". The clear rejection of any production only evolves in later discussions of play. Caillois' differentiates play from work by declaring it the "pure waste: waste of time, energy, ingenuity, skill, and often of money" (here cited from Salen and Zimmerman 2004).

Yet, the connection between play and craft can be drawn from Huizinga himself. He lauds the role of social play as adding cultural value pointing out that "play demands application, knowledge, skill, courage and strength" (Huizinga 1949). All of them are elements of crafting as well. And both, play and craft, include endless learning processes that allow for ever-improved performance.

A second and more layered connection is in the point of material interest. Crafting as a practice does not depend on "material interest" or profit making. No doubt, commercial interests are important for craft practices just as they have also shaped art and design practices. Crafters need to sell their goods to make a living— so do artists and designers. Practicing craft as skilled handiwork does not depend per definition on commerce. It focuses on the activity of making as productive engagement. As crafting includes countless failures before mastery is achieved, the commercial outcome of a craft exercise is not always the ultimate achievement, but the process remains central.

Likewise, Huizinga's definition of play is process-based and less concerned about a definition of the vessel than about the cultural significance. Definitions for games are most helpful (see e.g., Juul 2003) but they focus on only one possible option of play "as a special form of activity, as a 'significant form', as a social function" (Huizinga 1949). While play is more of a cultural communication form in Huizinga's view, it is connected to designed rule-defined vessels (games) in its application to game studies, which leads it away from its interpretation in regard to other possible vessels. One of those alternatives is play as craft. What if the often-separated elements of play as self-expression (play as performance) and play as fun activity (play as unproductive special state) and productive play (play as craft practice) are all part of that interwoven net formed by play and culture that Huizinga investigates?

4 Productivity and Critique

Different forms of production have been instrumental in the shaping of craft's history. One key development is the separation of production into hand- and machine-made, which was not existent before the earliest forms of industrialization. In his critical review of the shift, Adamson sees craft as "both a necessity and a problem for modernity" (Adamson 2013). He rejects a simpler view of craft as an antidote to modernity's embrace of new technology, a view that prominently shines through in statements from Morris to Sennett. For Adamson, craft is skillful practice but not necessarily the paragon of creativity as the most repetitive production of goods of equal quality is the basis of a craftsperson's work. While he acknowledges and discusses the practice of studio crafts that rely largely on individual vision and single piece production, he sides with craft as aimed for production. Others, like Shiner, discuss craft in an art historical context (Shiner 2010), separating it from the fine arts and their more conceptually driven agenda.

The relationship of craft to production is as dynamic as that of play to production, and it remains contested today. For millennia, skilled handiwork was the only way to produce necessary goods. Today, it is much more laborious, expensive, and not always beneficial to the quality of an object if we produce it by ourselves. Yet, it is a growing practice. The maker culture grows not out of economic need (even though times of recessions might spur more DIY) but as a form of self-expression and/or self-empowerment. In this regard, the modern revival of craft practices and its siblings in the DIY and maker movement not necessarily present a realistic revolt against the machines but an idealistic and often highly personal one.

Today's turn to craft includes many shades of motivation, from the anarchistic arguments of Cody Wilson, publisher of the 3D printable gun, to attacks on walled garden technologies and reclaiming of black-boxed technology such as jailbreaking iPhones, to the success of commercial outlets such as MAKE magazine and etsy.com, to Fab Labs and Open Source Community projects. Throughout, one can trace both a sense of reclaim of the practice (not necessarily against the industry but as a kind of technological self-liberation attempt) and the role of this reclaim in shaping communities and cultural development. As Greer argues: "We are no longer knitting together out of a necessity for basic clothing needs, we are creating together out of necessity for closeness" (Greer 2004). The need for closeness was also given in times when knitting was the only way to produce clothing, but society's focus has shifted and the revival of craft is offered as a cultural response to alienation and disconnect from social and material worlds. This relates back to Huizinga's "social function" of play and the particular form of "social play" (Huizinga 1949) that is a strong community building activity. Holding his critical stance, Adamson acknowledges the parallels between craft and storytelling, which we can see as origins for social construction, but he continues this argument toward a traumatic role of craft in the craft revival movement that uses the concept of memory. "This memory work is escapist, but not only that; it also describes modern experience in such a way that it can be critically engaged" (Adamson 2013). Adamson remains skeptic of the way craft is being adopted by digital communities as well as of the craft revivalist's approach. At the same time, his notion of craft as a critical approach emphasizes craft as a form of critical and practical engagement. The importance of critical thinking and practice will be the focus of the conclusion of this chapter.

Approaching play as productive activity, we arrive at a crossroads of craft and play not only in terms of practice but also as critical approaches. The notion of craft, in its social and productive practice, is added here to allow a new perspective to the making and the results of this practice. One that is not bound to pre-designed routes given in the original piece, such as a virtual or real economy, or to the solution of a pre-existent puzzle, but one that is unpredictable and creative at the same time. It might be repetitive but in the repetition lays a process of learning and continuous optimization.

5 Machinima

Machinima is the practice of using video games and other real-time virtual environments as stages for the production of moving image pieces. The results are most often videos recorded in the game engine and re-assembled to tell some narrative or to simply show some game action.

The lineage of machinima from video game play to production has been outlined before (Marino 2004; Lowood 2008; Kelland et al. 2005 for multiple critical perspectives see also Lowood and Nitsche 2011). Although these scholars focus on different perspectives, they retell the evolution of gamers from players to producers. Along this evolution, the players' involvement in and knowledge of the game change, grow, and are reframed. These changes open up the connection between machinima and craft as the technically defined play practice is opened up. Oakeshutt differentiated knowledge into technological and practical/traditional knowledge (Oakeshott 1962):

> The first sort of knowledge I will call technical knowledge or knowledge of technique. In every art and science, and in every practical activity, a technique is involved. In many activities this technical knowledge is formulated into rules which are, or may be, deliberately learned, remembered, and as we say, put into practice; but whether or not it is, or has been, precisely formulated, its chief characteristic is that it is susceptible of precise formulation. [...] The second sort of knowledge I will call practical, because it exists only in use, is not reflective and (unlike technique) cannot be formulated in rules. This does not mean, however, that it is an esoteric sort of knowledge. It only means that the method by which it may be shared and becomes common knowledge is not the method of formulated doctrine (Oakeshott 1962).

This distinction has also been applied to craft, where the concept of "tacit knowledge" as a form of practical knowledge is often cited as a typical sign of a crafter's expertise that cannot be properly documented. Technical knowledge is the defined rule system of a certain practice, but "tacit knowledge is learned and absorbed by individuals through practice and from other people; it cannot usually be learned from books" (Dormer 1997). In practice, tacit craft knowledge includes the awareness of how to handle certain materials or tools. For example, a potter might feel the difference of particular clay and thus form an object differently.

Expert players of video games display both kinds of knowledge: They have fully understood the underlying rules and conditions of the game system, but they also have developed strategies and techniques that often exceed the originally implemented design through forms of "emergent play" (Salen and Zimmerman 2004). Games can support such a development through their design (see e.g., Juul 2002 who distinguishes between open and closed game designs). However, it falls onto the players to exploit, share, and develop the opportunities that are provided.

The origins of machinima trace back to game companies and expert hackers, but as this argument looks into play as crafting, it will concentrate on the player-production. In many ways, player-produced machinima is the practice as well as documentation of emergent play. Their creative production depends on the technical frame as well as emergent practice of machinima production.

When the *Quake* clan The Rangers realized that they could use the demo recording ability of id's *Quake* game engine not only to document their playing skills and strategies but also to tell stories, they had amassed enough practical knowledge about the game to turn it from a closed technical system—a world to play in—into a tool—a world to play with. They used this tool to produce what is widely seen as the first player-produced narrative machinima: *Diary of a Camper* (1996). *Diary of a Camper* tells the story of a group of *Quake* avatars facing an unknown player in a particular part of a game map called The Dark Zone. They manage to eliminate this character only to identify him as John Carmack, co-creator of the game itself. The story is simple, the rudimentary dialogue is presented in pop-up text lines, and the visual quality of the textures and screen resolution are limited to the engine's performance. Yet, by performing a simple sketch within the *Quake* level, assigning each other roles, and recording this activity in the *Quake*-specific demo format, The Rangers transformed themselves from players into producers and crafters. The resulting object was the demo file, a single log file that allowed other players to share it, download it, load it into their copies of *Quake,* and witness the event being re-performed in the engine. The demo format still depended on the game engine to run. It, thus, was limited to the gamer community. However, later machinima productions—starting with Tritin's *Quad God* (2000) made in the *Quake III* engine—utilized screen capture methods that resulted in independent artifacts in the form of video files to be shared and distributed over the Internet. It is only fitting that *Quad God* was released on the just launched machinima.com web site, which would turn into an important portal for machinima creators and aficionados for the coming years. Later machinima work turned the game further and further into a creative tool with little limitations to what could be generated with it and it transformed play into using this tool—an activity we might also call crafting. What is particular for the world of machinima is the fact that this crafting also entails artistic expression. Despite all its simplicity, *Diary of a Camper* still displays an artistic vision and perspective that is performed by expert players.

> Recasting the player as a performer settled into not one, but two predominant modes: the superior player, the God of the joystick and mouse, and the player-programmer able to hack into game code and show off mastery of the technology (Lowood 2008).

Diary of a Camper adds a third category: that of the actor who is not performing the acquired mastery of the in-game logic but utilizes this knowledge to perform on top of it and for external, not-game-related means. In that case, machinima has become a performance platform not for pre-designed play but for free speech and expression (Carroll and Cameron 2005). And as in craft, limitation, personal expression, and production are heavily intertwined. This becomes obvious in the content choice of *Diary of a Camper*, which is not untypical for machinima.

Machinima shows a self-reflexive quality through its choice of topics and form of storytelling that engages with its game origins. This refers to the underlying game systems as cultural conditions, as seen in the most popular machinima series *Red versus Blue* (2003–) by Roosterteeth, which starts off as a comical interrogation of its original game setting in the game universe of *Halo*. It also includes the crafting process of content and play behavior itself, as seen in Robert Stoneman's *War of the Servers* (2007) produced in the *Half-Life 2* engine within the particular game mod *Garry's Mod*. The film is based on the concept album *Jeff Wayne's Musical Version of the War of the Worlds* (1978), which itself is a reinterpretation of the H.G. Wells original story from 1898.

In the case of *War of the Servers*, the focus on content creation and enhanced functionality of the underlying game options in *Garry's Mod* become an integral part of the machinima's storytelling: Players set out to defend their servers against an AI invasion. The two roles of player as "God of the joystick" and "player-programmer" are ultimately conflated in the story line of *War of the Servers*.

The result is an independent object, a downloadable movie file available at Stoneman's Web site (http://www.robertstoneman.com/wots/). *War of the Servers* not only displays the productive energies of the machinima makers, but it also celebrates the engagement with the game system. The game is the tool for production and simultaneously used as backdrop for the evolving events. This illustrates the playful reflection on gaming culture through the productive movement of playing as crafting.

On the one hand, machinima often operates as a technical celebration of the game's opportunities. A whole section of machinima deals with the documentation of glitches and exploits of games. There are numerous productions that concentrate on showing the game engine's performance such as Randal Glass's *Warthog Jump* (2002) produced in *Halo* as a "Halo Physics Experiment" to show that a combat vehicle could be "launched" into the air with the help of explosives. Like craft's focus on skills, this machinima focuses on the technical achievements of the underlying engines and the mastery of the "God of the joystick" who can make new sense of them (for more on players exploring game engines as productive tools, see chapter Little Big Learning in this volume).

On the other hand, the lines to the artistic performative are easily crossed. Randal Glass himself became an early contributor to the *Red versus Blue* franchise, playing a character in the episode *A Shadow of his Former Self* (2003). Technical knowledge of the game and its rule system, and the practical knowledge of emergent play that exceeds the restricted system at hand toward freer artistic expression go hand in hand. Through its combination of technical and practical knowledge and the inclusion of individual expression through practice machinima provides a useful example for playing as crafting. The merger of the player as "god of the joystick", player-programmer, and player-actor forms a fertile ground for the evolution of this craft-like practice.

6 Conclusion

Machinima is only one example where playing and crafting connect. To critically investigate this overlap, references are needed to both craft research and game studies. Not unlike the inclusion of other neighboring domains such as anthropology, film studies, gender studies, or performance studies, this enriches the scholarly palette with which we investigate video games. It shines new light on relevant issues such as skill, materiality, or economics and might also help us to re-investigate historical origins and future design paradigms. In the best possible case, play as craft leads to more questions and richer debates on games—and possibly to new challenges for craft as well.

Because this argument grew from the parallels of process-based playing and crafting, one challenge for both domains is the idea of critical production forms. "Critical play" has already found notable attention in game studies but largely as a form of artistic expression, not productive practice.

> Critical Play is built on the premise that, as with other media, games carry beliefs within their representation systems and mechanics. Artists using games as a medium of expression, then, manipulate elements common to games—representation systems and styles, rules of progress, codes of conduct, context of reception, winning and losing paradigms, ways of interacting in a game—for they are the material properties of games, much like marble and chisel or pen and ink bring with them their own intended possibilities, limitations, and conventions (Flanagan 2009).

Flanagan's perspective is helpful to define the critical component, and she certainly is keenly aware of the productive power of emergent play, but it is presented as a largely pre-structured criticism provided by the game designer as artist and to be realized in the playing of the resulting game through the player. The art is for the designer to "address intervention, disruption, and social issues and goals alongside of, or even as, design goals embedded into the mechanic and game elements" (Flanagan 2009). And the role of the player is to explore these mechanics.

In contrast, playing as crafting shifts the focus more on the idea of play as productive and critical at the same time and it emphasizes the role of the player. A useful reference point, then, is the concept of "critical making" defined as the use of "material forms of engagement with technologies to supplement and extend critical reflection and, in doing so, to reconnect our lived experiences with technologies to social and conceptual critique" (Ratto 2011). Critical making puts more emphasis on the making itself than on the particular outcome:

"The final prototypes are not intended to be displayed and to speak for themselves. Instead, they are considered a means to an end, and achieve value though the act of shared construction, joint conversation, and reflection" (Ratto 2011).

This seemingly clashes with the focus of this essay on productive play that generates some object forming beyond the production area it was generated in. Machinima was discussed as a kind of workbench that allows the production of independent objects, the animated films. It also steps away from craft's widespread perception as leading to particular object-outcomes whereby these objects might have own and unique qualities (Risatti 2007). Yet, while it challenges these perspectives, it does not contradict either of them. Critical making as an emerging scholarly field emphasizes the means of making as reflective and expressive activities that can help to form critique on the lived space of the producer. Machinima is one example where the materiality of the lived space, the space where shared production takes place, is digital. It can take the form of a *Quake* server and a shared game map in *Diary of Camper*, for example. This means that critical making can be part of emergent play within a largely digital condition. In the case of *Diary of a Camper* and *War of the Servers*, this is also notable in the continuous cross-referencing to underlying game culture that marks these machinima. We find this, for example, in the notion of one of the game creators as a "camper" in a *Quake* game level, or the countless references to server "admin" maintenance in *War of the Servers*. Critique, here, is mainly humorous—albeit far from apolitical. A third machinima example might clarify the idea of critical play as productive in more obvious ways.

Joseph DeLappe's *dead-in-iraq* (2006–11) was an online performance in the game *America's Army*, a game that was financed and distributed by the US Army. *dead-in-iraq* consisted of DeLappe logging into online game sessions, and instead of partaking in the pre-designed game behavior of shooting other players to win the game's round, he used the text-to-voice-chat to communicate the names of US soldiers fallen in Iraq to all other players present in the game session. The project is documented in extensive machinima recordings as well as screenshots of the unfolding results. For *dead-in-iraq,* DeLappe altered the pre-designed play into a combined practical critique of the material at hand. It addresses the game *America's Army* as a PR and recruitment tool for the US Army based on Epic's *Unreal* engine. It also is a response to the physical war activity in Iraq. DeLappe successfully uses the game as productive tool to critically interrogate both layers through play, realize the overlaps between them, and to critique them both. His play is not masterful in the sense of a "god of the joystick"—his game character usually dies very fast as he remains motionless in the game level and is an easy prey for the opposing team. But he mastered the game's system for his particular purpose as he uses the text-to-voice-chat that cannot be switched off by other players and thus provides an effective speaker's corner for his efforts. The piece works because it builds on the pre-designed intentions of the system at hand in an innovative way as it uses the game as a flexible set of tools for its expressive means. DeLappe creates, what he terms a "live, machinimatic act" (DeLappe 2013) in the tradition of interventionist performance art.

The final question, then, is how to design for such a craft-related and expressive tinkering with the existing game to encourage critical reflection through the practice of playing. Flanagan provides a framework for the design of critical play by including values into the design process and the evaluation during iteration (Flanagan 2009), and her practice exemplifies the collaborative creative production methods mentioned by Ratto. But the idea of play as craft suggests to focus less on the ways that designers can translate their vision, values, and artistic concepts into the game system and instead to approach games as tools for players to create and express their own positions and craft representative objects through them. Crafting/ playing is a critical production process. Ultimately, it is another trace of cultural production through games and positions them in relation to new territory and new questions.

References

Adamson, G. (2013). *The invention of craft*. London, New Delhi, New York, Sydney: Bloomsbury.

Ahn, L. V., Dabbish, L. (2004). Labeling images with a computer game. In *Proceedings of the SIGCHI Conference on Human Factors in Computing Systems* (pp. 319–326). Vienna, Austria: ACM.

Bonanni, L., Parkes, A., Ishii, H. (2008). Future craft: how digital media is transforming product design. In *CHI '08 extended abstracts on human factors in computing systems* (pp. 2553– 2564). New York: ACM.

Bunnell, K. (2004). *Craft and digital technology*. Presented at World Crafts Council. Metsovo, Greece.

Camper, B. (2005). *Homebrew and the social construction of gaming. community, creativity, and legal context of amateur game boy advance development*. M.Sci. thesis in Media Studies MIT, Cambridge, MA.

Carroll, J., & Cameron, D. (2005). Machinima: Digital performance and emergent authorship. In *Digra 2005: changing views—worlds in play*. Vancouver, CAN.

Castronova, E. (2005). *Synthetic worlds: The business and culture of online games*. London, Chicago, IL: The University of Chicago Press.

DeLappe, J. (2013). Playing politics: Machinima as live performance and document. In J. Ng (Ed.), *Understanding machinima. Essay on filmmaking in virtual worlds* (pp. 147–167). London, New Delhi, New York, Sydney: Bloomsbury.

Dormer, P. (1997). Craft and the turing test for practical thinking. In P. Dormer (Ed.) *The culture of craft* (pp. 137–158). Manchester, UK/ New York: Manchester University Press.

Flanagan, M. (2009). *Critical play: radical game design*. Cambridge, MA, London: MIT Press.

Gershenfeld, N. (2012). How to make almost anything: the digital fabrication revolution. *Foreign Affairs, 91*, 43–57.

Greer, B. (2004). *Taking back the knit: Creating communities via needlecraft, in culture, globalisation and the city*. M.A. thesis, Goldsmiths College, London, UK.

Huizinga, J. (1949). *Homo Ludens: A study of the play-element in culture*. London, Boston, MA, Henley: Routledge & Kegan Paul.

Juul, J. (2002). The open and the closed: Games of emergence and games of progression. In F. Mäyrä (Ed.), *Computer game and digital cultures* (pp. 323–329). Tampere, FI: Tampere University Press.

Juul, J. (2003). The game, the player, the world: Looking for a heart of gameness. In M. Copier, & J. Raessens, (Eds.), *Level Up: Digital Games Research Conference* (pp. 30–45). Utrecht, NL: Utrecht University.

Kelland, M., Morris, D., & Lloyd, D. (2005). *Machinima*. Boston, MA: Thomson.

Khatib, F., et al. (2011). Crystal structure of a monomeric retroviral protease solved by protein folding game players. *Nature Structural & Molecular Biology, 18*, 1175–1177.

Lastowka, G. (2009). Rules of Play. *Games and Culture, 4*(4), 379–395.

Lindtner, S., et al. (2009). Situating productive play: Online gaming practices and Guanxi in China. In *INTERACT* (pp. 328–341). Heidelberg: Springer.

Lowood, H. (2008). Found technology: players as innovators in the making of machinima. In T. McPherson (Ed.), *Digital youth, innovation, and the unexpected* (pp. 165–196). Cambridge, MA: MIT Press.

Lowood, H., & Nitsche, M. (Eds.). (2011). *The machinima reader*. Cambridge, MA: MIT Press.

Marino, P. (2004). *3D game-based filmmaking: the art of machinima*. Scottsdale, AZ: Paraglyph Press.

Masterton, D. (2007). *Deconstructing the digital*. Presented at: New Craft-Future Voices (pp. 7–24). Dundee University.

McCullough, M. (1998). *Abstracting craft: The practiced digital hand*. Cambridge, MA: MIT Press.

Oakeshott, M. (1962). *Rationalism in politics and other essays*. London, New York: Methuen/Barnes & Noble Books.

Pearce, C. (2006). Productive play. game culture from the bottom up. *Games and Culture, 1*(1), 17–24.

Ratto, M. (2011). Critical making: Conceptual and material studies in technology and social life. *The Information Society: An International Journal, 27*(4), 252–260.

Risatti, H. (2007). *A theory of craft. Function and aesthetic expression*. Chapel Hill: University of North Carolina.

Salen, K., & Zimmerman, E. (2004). *Rules of play: Game design fundamentals*. Cambridge, MA: MIT Press.

Shiner, L. (2010). The fate of craft. In S. Alfoldy (Ed.), *NeoCraft: Modernity and the crafts* (pp. 33–47). Halifax: The Press of the Nova Scotia College of Art and Design.

Taylor, T. L. (2006). Play between Worlds. *Exploring online game culture*. Cambridge, MA, London: MIT Press.

Author Biography

Michael Nitsche is associate professor in digital media at the Georgia Institute of Technology where he teaches mainly on issues of hybrid spaces and what we do in them. He uses performance studies, craft research, HCI, and media studies as critical approaches and explores this borderline in digital media, mobile technology, and digital performances. He directs the Digital World and Image Group and is involved with various interdisciplinary research centers.

Michael's publications include the books *Video Game Spaces* (Cambridge, 2009) and *The Machinima Reader* (Cambridge, 2011; coedited with Henry Lowood, both MIT Press).

Playmakers in the Maldives

Amani Naseem, William Drew, Viktor Bedö and Sidsel Hermansen

Abstract This chapter describes the Playmakers in the Maldives project, a collaborative work made for the Maldives Exodus Caravan Show at the 55th Venice Biennale. The project involved ten international game designers collaborating with the Maldivian communities and individuals to create games and play events in the public spaces of the Maldivian capital island Malé. We present the games developed and describe some of the issues that we faced during the project.

> Playing a game in public is a political act.
>
> —DeKoven (2013a)

This chapter describes Playmakers in the Maldives (April–June 2013), a project where ten designers from international play communities visited the Maldives to create games for public spaces in the capital island of Malé in collaboration with local creative communities. During the project a total of seven games were created and played in Malé. A number of small semi-public events and a larger final public event were held where the new games were developed and played, as well as a number of other more established street games, for example, Turtle Wushu by Invisible Playground. This project was a collaborative work made for the 55th Venice Biennale. Games from the project were shown and played at the Museum of

A. Naseem (✉)
Copenhagen Game Collective, Copenhagen, Denmark
e-mail: amaanii@gmail.com

W. Drew
London, UK

V. Bedö
Tacit Dimension, HPI School for Design Thinking, Potsdam, Germany
e-mail: viktor@tacitdimension.com

S. Hermansen
Copenhagen, Denmark

© Springer Nature Singapore Pte Ltd. 2018
D. Cermak-Sassenrath (ed.), *Playful Disruption of Digital Media*,
Gaming Media and Social Effects, https://doi.org/10.1007/978-981-10-1891-6_8

Fig. 1 Malé, the capital island of the Maldives

Everything official Collateral Exhibition as part of the Maldives Exodus Caravan Show.

By making games to be played in public the Playmakers in the Maldives events highlighted the precariousness of public space in the Maldives, particularly in the nation's capital, Malé. In the past decades, the Maldives—like many countries in the developing world—has experienced rapid urbanization, which has had a tremendous impact on the capital Malé (see Fig. 1). As a consequence, the Malé streets have changed dramatically, becoming congested with heavy traffic. New housing developments have replaced traditional Maldivian houses that were open to the street. Space is a premium, and neighborhood play in public has become virtually non-existent. The scarcity of public space caused by urbanization is compounded by the impact of climate change on the island nation. The Maldives is one of the lowest lying countries in the world placing it in immediate danger from rising sea levels, and the small islands of moving coral sand are extremely vulnerable to any climate change or shift in ocean currents. Public space in the Maldives is thus also threatened by global climate change making its existence even more precarious in the island nation.

Furthermore, the Playmakers project took place at a time of public turmoil in the Maldives, following a military coup on February 8, 2012, that toppled the country's first democratically elected government. In the wake of the coup, the public spaces of Malé, especially the street life, became very politicized. In the months since the coup, there had been numerous demonstrations and brutal police crackdowns, and a law had been passed making it illegal to have more than three people together in

Fig. 2 Trying out the game "Once Upon a Time... Bread" by Lena Mech on Villingili, a suburb island of Malé. Image copyright Hussain Asthar

public without permission from the government. Creative expression had also become very politicized and polarized, the year after the coup saw a huge burst of creative output, mostly critical of the coup, in music, art, and media from practitioners in the Maldives. In part, protest was channeled into these events because the ban on assembly exempted 'cultural events' and 'games,' although the period also saw an unprecedented number of huge political rallies and demonstrations. In this context, the goal of the Playmakers project to bring play to the streets was perplexing, daunting, and challenging.

This chapter describes our activities in this challenging situation, where urbanization, climate change, and political oppression combined to put the future of public space and public play in the Maldives in question. It will present the games developed and how they were played in Malé, how we created situations for play and invited people to join in (Fig. 2).

1 Background

This project was a collaborative work between an international group of artists and game designers. We were Andrea Hasselager (Play and Grow), Amani Naseem (Copenhagen Game Collective), Patrick Jarnfelt (Copenhagen Game Collective), Ida Toft (Copenhagen Game Collective), Sidsel Hermansen (Copenhagen), Kunal

Gupta (Babycastles), Joseph Ahearn (Silent Barn), Lena Mech (I Spy), Viktor Bedö (Invisible Playground), and William Drew (Venice as a Dolphin). From our group, Amani Naseem had been invited to show at the Venice Biennale, and being a Maldivian involved in growing European movements for public play, the idea to take play to the streets in the Maldives was always there. The Venice Biennale exhibition provided the impetus to finally organize a project that would bring play to the Malé, while the aftermath of the military coup gave the project a certain urgency. It was difficult to say as the plans for the project took place, what the situation on the ground would be from one day to the next, but all the designers agreed to go and try to do our work anyway.

The theme for the Maldives Exodus Caravan Show at the Venice Biennale was to expand conversations around issues of climate change and the cultural and political unrest in the Maldives. Former president, Mohamed Nasheed, holding an underwater cabinet meeting (Fig. 3), shows an example of the playful side of Maldivian culture. As we found in discussions during our time there, humor and playful approaches are often adopted to cope with or tackle serious issues. This was reflected in the creative explosion following the coup, where writers, artists, and musicians were extremely productive, using social media and the Internets in an unprecedented way to reach an increasingly connected population.

Play has been regarded as central to and a source of culture (Huizinga 1992), an important way of being together (De Koven 2013b), crucial to humans' ability for adaptability and creativity (Brown 2008), a form of human expression and

Fig. 3 Maldives Cabinet Dive organized by Divers Association of Maldives. Image copyright Mohamed Seeneen

representation (Sutton-Smith 1997), a way of creatively engaging with the world (Sicart 2011). This points to the potential that games and play can have in the conversations and exchanges around futures of cities and communities. In our case, we wanted to engage in the Maldivian situation through playing in public and creating games for the streets of Malé.

Play in the modern city is normally compartmentalized, occurring in defined spaces such as arenas, gyms and playing fields, and at defined times, often separate from work and other daily activities (Eichberg and Nørgaard 2005). In Malé, with the overcrowding and lack of space, even these designated spaces are few and far between. The intention was to create new avenues for public play in Male, as we might have done in any city. The coup, the resulting polarization of the Malé community and the fraught atmosphere made finding different ways of being in public spaces and starting conversations even more important. Theorists from various disciplines have made explicit links between the ideals of togetherness as central and necessary to the activities of play, game, and festivity (Wilson 2011). In the Maldives, facing climate disaster as well as political upheaval, bringing games to the public spaces was about introducing playing as a possible way of coming together in public spaces, and within the Maldives Caravan Show project, as a possible way of starting conversations and of public expression.

2 Conversations and Collaborations

The designers who visited the Maldives were all actively involved in creating play communities and cultures around games and play in their own countries both as individuals and as part of companies and collectives. This was not only an opportunity to play and create games in a new environment; it was also an opportunity to introduce the playful culture scene to people who had not encountered it before. We did not just want to present our own games; the intention was also to start conversations, opening up possibilities, and creating space for collaboration. We wanted to present our previous work, and play games, but also show the possibility for people to create their own games to play in public spaces (Fig. 4).

We announced a number of small gatherings in café's or other public locations to talk about our project and play games as a way of introducing ourselves and getting to know interested people in Malé. Most of these events were small meetings, among groups of friends who already knew each other. Later on in the week, we would often split up and go and meet people individually, such as musicians, sports people, artists, DIY activists, political activists, and writers (Fig. 5).

We would have liked to collaborate in creating the games themselves, but without formal structures, this was not possible given the short period of time we were there. On the other hand, a number of people would regularly join in to try out games we were developing and discuss how we were making them. The coup and the fact that a lot of people were heavily involved in political work in addition to

Fig. 4 Meeting with the DIY environmentalist group Villimalé Beach Clean Up at their compound in Villingilli, an island suburb of Malé. Image credit Ida Marie Toft

Fig. 5 Democracy activists Susan Ibrahim (with brainwave monitor) and Marie Zahir, and artist Mariyam Omar with curator Elena Gilbert trying out the game Idiots Attack the Top Noodle at a café in Villingilli, a suburb of Malé. Image copyright Hussain Asthar

Fig. 6 Playing the Danish traditional game Telephone on a boat play session. Elena Gilbert with Vaka, our food sponsor from Top Up and Mushfique Mohamed a democracy activist and journalist Aina Hannan. Image credit Ida Marie Toft

their regular jobs also meant that people did not have the time to dedicate to work with us for long periods every day (Fig. 6).

Often people were concerned about and reluctant to meet because the gatherings may be seen as political and shut down by or worse arrested by the police. Indeed, conversations often turned to political issues and stories. The lack of freedom of expression in public was a huge concern for most, having experienced police violence during demonstrations and persecution of journalists. As designers, this became a central concern when creating games and gathering for events, as we did not want to endanger players. Another difficulty of gathering people together was that some would not join thinking that these were political gatherings, while others were reluctant because they were unsure as to which side we were taking. Trying to keep the gatherings as open as possible was not defined enough for the polarized public.

3 The Games

This section describes the games developed during the project. These games were played during our stay and for the Maaja Making event at the end of our project. Some of the games were also played during the Venice Biennale opening week at the Maldives Exodus Caravan Show.

4 The Hunt for the Yellow Banana

The Hunt for the Yellow Banana was a city-wide cross-platform game story that took place in Malé in April 2013. It began with the distribution of flyers in English and Dhivehi (Maldivian language) and the putting up of posters announcing that a Banana had gone missing. On the flyers and posters was a photograph of an identifiable human wearing a banana outfit with sunglasses on and an e-mail address and phone number to contact in case anyone spots the Banana (Figs. 7 and 8).

Members of the team (principally the game's core design team Andrea Hasselager and William Drew) would then appear as the Banana in different parts of the city. There was a Web site with a map that would mark sightings as well as a Twitter hashtag. In the lead up to the final live event, which was a team competition to find the Banana, the narrative took a turn whereby the Banana, alone in the city and having discovered its autonomy, was feeling depressed and had been contemplating suicide. There were various clues posted online as to how the Banana might choose to end its fruitful existence on Earth. After the designers flew back home, the Banana inexplicably made simultaneous appearances in both Malé and Venice. Since then there have been occasional sightings in bars and clubs in Copenhagen and at Roskilde Festival, though the jury is still out on whether these are original Banana appearances or copycats.

5 Maldives Trading

This was a game by Kunal Gupta and Joe Ahearn that took place for most of the duration of our stay. Players traded objects and their stories in an alternate economy based on the value of the objects stories. During the project, the stories were recorded in the trading book that was placed at one of the surfer hangout spots by the ocean in Malé. The game started with the designers trading their own artworks. The game might still continue, with the art works being traded and passed from hand to hand somewhere in the Maldives or even abroad. One of the original art works has been spotted in the Netherlands (Fig. 9).

6 Dreams in a Bottle

Dreams in a bottle is a game by Lena Mech. She asked a number of people to write down dreams that they wanted to be fulfilled and put the pieces of paper in bottles. The bottles were consequently passed from person to person. In order to pass a bottle to another person, players would have to choose a dream from the bottle and

Fig. 7 William Drew as The Banana on a scooter ride through Malé. Image copyright Hussain Asthar

Fig. 8 Flyer for The Hunt for the Missing Banana on a Malé street wall. Image copyright Hussain Asthar

fulfill it for the person they were giving the bottle to. Once they had done that they could add their own dream into the bottle and pass it to the person whose dream they had fulfilled (Fig. 10).

Fig. 9 Joseph Ahearn and Kunal Gupta introduce their game Maldives Trading at the Maaja Making event. Image copyright Hussain Asthar

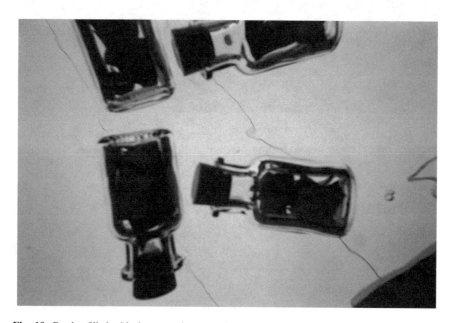

Fig. 10 Bottles filled with dreams written on pieces of paper. Image credit Lena Mech

7 Bite-Sized Water Games—Jelly Stomp and Poison Sea

These games were made by Sidsel Hermansen, Patrick Jarnfelt, Ida Toft, and Amani Naseem and use waterproofed PlayStation Move controllers. These were simple physical games, one for playing in shallow water and one for deeper water. The games used the break in Bluetooth connectivity caused when the floating controllers dipped underwater. Each controller could then have a number of lives.

Jelly Stomp works in a similar way to the common children's party game Balloon Stomp, and the controllers are attached to the players' wrists with elastic. It can be played in teams or one on one, where opposing players try to kill off each others "Jelly" by stomping on them in shallow water (Figs. 11 and 12).

Poison Sea is played in a demarcated area with controllers placed around it. A Conservationist and a Resort Tycoon protect or try to kill the jellies. The Conservationist, who has developed immunity to the Jellies protective poison, can agitate the jellies when the Resort Tycoon is near to win by poisoning the Resort Tycoon. The Resort Tycoon wins by eliminating all the Jellies before the Conservationist revives any of them.

Fig. 11 Playing Jelly Stomp at Artificial Beach during the Maaja Making event. Image credit Sidsel Hermansen

Fig. 12 PlayStation Move controllers waterproofed with condoms and sticky tape. Image credit Ida Marie Toft

8 Operation Noose

Story: An intruder landed in your country. She has been located in your district and is on the run now. It is up to you and your special force buddies to arrest her by tightening the noose and closing the trap. But be aware! The intruder has tapped into your communication network. She knows what you know. She can outmaneuver your tracing equipment and render it less accurate. Can you and your special force buddies prevent the intruder from escaping by using group strategy, intuition and local knowledge?

(description from http://tacitdimension.com/portfolio/operation-noose/)

Operation Noose, is a smartphone-based chasing game and it was play tested a number of times during our project, and publicly played during the Maaja Making event. Three to five Special Force Teams had 20 min to capture an intruder on a playing field, the size approximately six by six streets (Fig. 13).

The play experience is based on the vague difference of *place* and *location* using the messaging mobile app—WhatsApp. When sharing location, other users instantly receive the position on map based on GPS coordinates. When sharing place, the user chooses from a scroll down menu containing nearby places like cafes, hospitals, schools, and other places that are in the cartographic database that WhatsApp is using. So when the intruder shares her place during the game, she can tactically choose places in a quite big radius. In Operation Noose, Special Forces

Fig. 13 Chef Jailum Hameed checks his phone during a round of Operation Noose in Malé. Image credit Viktor Bedö

request the position of the intruder by disclosing their own location. The Special Forces can request the location only twice in one game, all the other times they request the place of the intruder. As a result, the playing experience is less about following a clear trail and then about vague directional pattern in an economy of accuracy.

Operation Noose challenges the overlap between cartographic databases that are fed into mobile apps and the actual experience of urban space in several ways. The right choice of the size of the playing field is a key to balancing out the chance to win for the teams. The optimal size not only depends of the number of team but also on unique features of the neighborhood that are not or only partly represented through map data. So the choice involves local knowledge about the area. Also when they are offered 'nearby places' by mobile apps, players need to develop an understanding of how close these places really are.

9   The Goat Herder and the Fainting Goats

This game is inspired by a story that was spreading in Malé about a group of drug smugglers who had close connections with government officials. The designers were Lena Mech, Amani Naseem, Patrick Jarnfelt, Ida Toft, and Andrea Hasselager (Fig. 14).

Fig. 14 Setting up for a game of Goat Herder and the Fainting Goats. Image copyright Hussain Asthar

A goat herder tries to smuggle drugs to a safe point using his herd of goats, while the police, with their drone, try to obtain evidence of the smuggling activity.

The game is played in three teams, the goat herder, the goats, and the police. The goats wander around the play area, one of them carrying 'the drugs'. They are slightly suicidal and change their direction randomly with every three steps. The goat herder has to direct them by standing in front of each goat and positioning her arms. Further, the goat herder has to make sure that he or she transfers the drugs from goat to goat in time so that no one goat dies from overexposure (timed) to the drugs. Meanwhile, the police have to direct the surveillance drone using their lights (PS Move controllers) that have to be periodically charged at charging stations (magnets placed at the edges of the play area). The police have to direct the drone toward a goat that is carrying drugs and take a picture with the drone to get evidence to win the game and arrest the goat herder before she reaches her safe point.

This game was played in Malé as the last game of the Maaja Making event.

10 Maaja Making

The main collaborative work turned out to be "Maaja Making", the final event we made at the end of our project. Artists and musicians joined us to create a public event for dancing, play, live art, and exhibition. Many were involved in preparations and helping out on the day. All the artists were present, and unlike a "normal" exhibition, the artists were having conversations about their work with the public, in the same way as we were presenting our games in person (Figs. 15 and 16).

Fig. 15 A small group gathers as the music starts at Raalhugandu during the Maaja Making event. Image copyright Hussain Asthar

Fig. 16 Kunal Gupta with artworks for the Maldives Trading game at the Raalhugandu area during the Maaja Making event. Image copyright Hussain Asthar

Maaja is a miss-spelling of the Maldivian word "Majaa"—"fun," a mistake by Elena the curator that added to the mix of cultures present. The event was held in April 2013 in the Raalhugandu area of Malé, an open space near the local surf point. All the designers presented their games with local artists and musicians who showed their work and played music outdoors (Fig. 17).

Although Maldivian law states that "cultural events" and "games" can be held without permission, our applications for permission were being repeatedly rejected. We tried to get permission for several different locations and times without success. Many of the Maldivian artists and the visiting designers were concerned about creating problems with the authorities if we held a large public event without permission, and it looked like we may not be able to continue (Fig. 18).

After much discussion with our collaborators, we decided to try again, this time presenting the event as an "art exhibition." Elena Gilbert, one of the curators of the Maldives Exodus Caravan Show and a collaborator on our project, had to officially take responsibility as the main organizer of the event. This was because unlike a play event, an art exhibition is not an unusual occurrence, and while previous applications to hold a "games and play event" were declined, the art exhibition curated by an international curator was approved (Fig. 19).

A popular local pass-time is going for a motorbike ride in the evening. Hundreds of people stopped by the event area and rows and rows of people watching on parked motorbikes causing a traffic jam in the area (Fig. 20).

Fig. 17 Journalist Aina Hannan and Democracy activist Mushfique Mohamed watch as members of The Underground Movement prepare the music set up for the Maaja Making event. Image credit Hussain Asthar

The military police then showed up in the middle of our playing area and demanded to know what was going on. We told them that we had written permission and Hussain Asthar, a photographer, rushed home to fetch the letter. But as soon as the police arrived, people crowded around and small arguments started. A lot of the people there had recently experienced clashes with police, during demonstrations, and some had even been detained arbitrarily. The police strode around threateningly, scaring some people; this exacerbated the existing tensions and could have escalated easily to a confrontation (Fig. 21).

We needed to create a diversion to diffuse the situation. Suggesting that they be careful not to scare the participants, we drew the police to a side. They did not want to play a game, but luckily directing the police to Elena the curator, who obliged them with detailed explanations of each exhibited artwork, occupied them during the wait for the permission letter. By the time the permission letter was presented to the police, they were more than happy to leave quickly, and unfortunately, they were no longer interested in the artworks on display (Fig. 22).

There were hundreds who came to watch, causing the traffic jam, but only a few dozens of people came to be actively involved in the event. We tried to invite people parked on their motorcycles to join in, but they would either drive off or refuse. Unlike the locations of our smaller meetings and events, the Raalhugandu

Fig. 18 Curator Elena Gilbert from the Maldives Exodus Caravan Show with a toy gun and a cheese sandwich during the Playmakers project. Image credit Rahil Naseem

area is a very open and visible. Also, unlike the previous events, this event featured live musicians with a really loud sound system. Although we invited people to come and see the artwork on display, people declined. Without explicit political affiliations, as an event during the period of the coup, it was an anomaly and hence could have been perceived as risky to be involved in so visibly.

There had recently been letters circulated by some religious groups against "mixed gender" dancing. This was a story that had been brought up by many of the young people we met that could have added to the lack of dancers that evening. There were small underground movements of people who would make parties outside Malé where it was more private, and huge political rallies with music and dancing were also common. Again, our event was outside any of these established events and therefore more difficult to join in. We were later told by some of the participants that there had been some people calling them "laadheenee" (irreligious) at the event, a political slur often used against the Maldivian Democratic Party. They said that a lot of the young participants had been really itching to join in and dance, but had felt inhibited by the amount of spectators. One of the participants was a fire dancer who made two performances to the music, and when she was dancing, a lot of the public would walk into the area to gather around and watch. When she was dancing, it was a more defined "performance" where the spectator and the audience were clearly separated from each other and it was easier for people

Fig. 19 Marine Biologist Rahil Naseem paints a fish during Sobah's graffiti workshop at Raalhugandu. Image copyright Hussain Asthar

Fig. 20 Hundreds of spectators sitting on parked motorcycles around the Maaja Making event area. Image credit Ida Marie Toft

Fig. 21 Amani Naseem introducing the Military Police to Elena Gilbert the curator to discuss the art on display. Image copyright Hussain Asthar

Fig. 22 Graffiti artist Sobah during his workshop at the Maaja Making event. Image copyright Hussain Asthar

Fig. 23 Artist Sumii Haleem, her partner airline steward Mohamed Almas, ex-intelligence analyst Nooshin Waheed and friends take a break after a game during the Maaja Making event. Image copyright Hussain Asthar

to gather around in the area. The moment she finished, people walked away and only watched from outside the area. This was unlike the games where the people gathering around could also easily become players themselves. Then, most people would stand far away and watch, the physical distance showing that they were not part of the happenings. Some of the Maldivian artists present also commented on this explaining that they had experienced this distancing and that a lot of people were scared to be actively involved because this was a new type of event, and political and social connections were not defined. At this point in time, in Malé, it seemed that our openness was daunting and made us less approachable (Fig. 23).

Similarly, more strangers from the public joined in the public play sessions on the beach in the nearby suburb of Villingili, where there was no passing traffic and less possibility for spectators to gather or police to appear. This could be why only a few people danced, while most sat along the seaside and watched on. It was the same for some of the games. In this context, the physicality and visibility of some of games we made became a barrier to joining in. Games like Jelly Stomp that involved splashing around in the sea, or Goat Herder and the Fainting Goats where some players made goat noises every once in a while, involves acting in awkward and silly ways.

The last game of the event was The Goat Herder and his Fainting Goats, after which we gathered with a smaller group of people who stayed on to ask further about our activities and also about making games. While the smaller events earlier had been mostly among groups who knew each other, the Maaja Making event attracted a broader group of people, a number of them who were interested in creating games themselves. This was a younger group who were educated and looking for new cultural activities and interested in being involved in future events. Since one of our goals was to introduce people to the possibility of making their own games, we were encouraged that a number of people suggested that they would like to participate in creating games themselves for a future event.

11 Venice Biennale

We played Jelly Stomp at a flooded St. Marks Square, Dreams in the Bottle was shown at the exhibition venue at the Sierra Dei Giardini, and the Banana also made appearances at the 55th Venice Biennale. It was extremely difficult to get players in Venice, during the busy opening week of the Biennale. Most of the people who exhibited at the Maaja Making event also joined us at the Maldives Exodus Caravan Show at the Venice Biennale, and the exhibition will be showing at different museums internationally in the next few years, giving us more excuses for further playful collaborations (Figs. 24 and 25).

Fig. 24 Playing Jelly Stomp in the flooded St. Marks Square during the Venice Biennale. Image credit Amani Naseem

Fig. 25 The Banana with a placard that reads "Down with this tipo di cosa" by the Serra Dei Giardini at the Venice Biennale. Image credit Bobbie Esra

12 Conclusion

Play can be a critique of existing conventions, a way of allowing people to "experiment with rules, roles and meanings" (Stevens 2007). The Playmakers in the Maldives project intervened through play by creating games and encouraging play in the public spaces of Malé. The games that were developed and playful practices that were encouraged challenged the politicization of public space through community re-appropriation of those spaces. The Hunt for the Yellow Banana engaged the public on many levels from making passing jokes to taking a picture, telling players where the Banana went, or joining in the hunt itself. The Banana was such a success that the suit was taken over by some of our collaborators in the Maldives. The Banana has since appeared on different occasions around Malé (Fig. 26).

The Bite-Sized Water Games were made to be played in the sea around Malé beyond the edges of the land. They used the unundation of the controllers as the central game mechanic. Where the game is played and the way it uses technology can be related to the precariousness of Malé's public space through potential flooding and the dangers of climate change. The events and games also illustrated how playful practices could reconnect people in public space and renew sociality in the streets of Malé. Maldives Trading, for example, created an alternate trading community centered around the swing at Raalhugandu surf point, where stories and personal conversations made up the value of game objects. For Dreams in a Bottle,

Fig. 26 A fruit seller gifts the Banana with a complimentary banana from his wares. Image copyright Hussain Asthar

players fulfilled each other's wishes in sometimes serious, but often very playful and hilarious ways during players everyday lives.

Our project is an example of how artists and game designers from different locations and cultural backgrounds can collaborate to develop public interventions that touch both global and local issues. Although our intention with the project was not to be subversive but to be as open and as possible, it ended up being so in some ways because of the situation around us. Play as an organizing concept provides a mode of engagement with and in public space that highlights that its role is not functional, but also generative of new relationships, ideas, and potential politics.

As a way to open up for conversations and collaboration for the Maldivian creative community, the project created international possibilities. Apart from the Venice Biennale, some of the local artists have been featured in international publications such as Art Asia Pacific Journal, among others, and the Maldives Exodus Caravan Show along with this project has been invited to a number of museums and galleries in Europe, the USA, and New Zealand in the coming years. All the designers are in touch with many individuals and groups in the Maldives; we are encouraged by the continuing conversations and interest in creating games and events together in the future.

Acknowledgements We thank all the designers and Maldivian collaborators for their support and for joining in the play. Special thanks to Mookai Suites and Top Up. Thanks to Thomas Apperley for feedback.

References

Brown, S. (2008). *Play is more than fun. TED talk. Serious Play*. Retrieved from http://www.ted.com/talks/stuart_brown_says_play_is_more_than_fun_it_s_vital.html.

DeKoven, B. (2013a). *IndieCade. A conversation with Eric Zimmerman & Bernie DeKoven*.

De Koven, B. (2013b). *The Well-Played Game*. The MIT Press, Cambridge, MA.

Eichberg, H., & Nørgaard, K. (2005). Education through play and game: Danish experiences. *International Journal of Eastern Sports & Physical Education, 3*(1), 166–197.

Huizinga, J. (1992). *Homo Ludens. A study of the play-element in culture*. Boston: Beacon Press.

Sicart, M. (2011, December). Against procedurality. *Games Studies: The International Journal of Computer Game Research, 11*(3).

Stevens, Q. (2007). *The Ludic City*. Routledge.

Sutton-Smith, B. (1997). Play as emotional survival. In *International Play Association Conference, Cardiff, UK*.

Wilson, D. (2011). *Brutally Unfair Tactics totally OK Now: On Self-effacing Games and Unachievements*. Games Studies, 11(3). http://gamestudies.org/1101/articles/Wilson

Wilson, D. (2012). *Designing for the pleasures of disputation-or-how to make friends by trying to kick them!* Ph.D. Dissertation, IT University of Copenhagen.

Author Biographies

Amani Naseem is a digital designer and researcher from the Maldives currently based in Copenhagen and Melbourne. She works in transdisciplinary groups to create experimental games that have been shown at international exhibitions and conferences and festivals. A member of the Copenhagen Game Collective, she has been involved in promoting game culture by organizing game jams and curating workshops and events among the indie games scene in Copenhagen. Amani cofounded and curated at w00t—Copenhagen Festival of Games and Play—and is currently a Ph.D. student at the Royal Melbourne Institute of Technology in Australia.

William Drew is a writer and a game designer based in London. He studied English literature at the University of Bristol and drama and theater at Royal Holloway, University of London, before spending several years working for the Royal Court Theatre. He started making interactive work in 2012 in Venice as a Dolphin and has since made games for a variety of contexts at festivals throughout the UK and Europe. He is a regular collaborator with Coney with whom he made Futureplay in 2013, an ongoing project in development. He was part of the inaugural Innovations in Storytelling program organized by the Stellar Network and the University of Roehampton. His writing about games has appeared in Kill Screen, *Wired*, Rock Paper Shotgun, and Unwinnable.

Viktor Bedö, Ph.D., is a street game designer and philosopher based in Berlin. He is the founder of the urban game design label Tacit Dimension and is a lecturer in design thinking. His activities are organized around his passion for urban dynamics, visual thinking, and team-based artistic and creative processes. In his Ph.D. thesis, "Interactive Urban Maps as Instruments of Thinking," he examined how bodily experience of urban life enables the discovery of emerging patterns on real-time urban maps. In Olafur Eliason's Institute für Raumexperimente, he cooperated with young artists on turning theoretical concepts of space and body into interactive performances. He is a former member of the street games collective Invisible Playground with whom he has been extensively touring international arts festivals (e.g., Copenhagen Art Festival, Oerol Festival Holland, Hide&Seek Weekender London). His recent game, Operation Noose, is part of The

Maldives Exodus Caravan Show that was featured by The Museum of Everything, a collateral program of the 2013 Venice Biennale.

Sidsel Marie Hermansen is a Copenhagen-based musician and sound and game designer. With a background from the vibrant alternative music scene of Copenhagen in several bands, and a recent Masters degree in game design from the IT University of Copenhagen, one of her main goals is to build bridges between the game scene and other artistic fields. Sound and music, and especially the gray areas in between, is a pivotal point of departure for most of her work, as is the urge to use game design in nonconventional contexts. This has led to her creating and curating experimental games for music festivals such as the Roskilde Festival and Henry's Dream, working with interactive fiction and creating interactive sound installations, as well as teaching experimental game design to students at Vallekilde Game Academy.

A Civilized Society

Eva Mattes and Franco Mattes

We hired several anonymous crowdsourcing workers through an online market-place and asked them to protest in front of their Webcams. We do not know who or where they are, and we could not ultimately predict what would result from our request. What came to fruition are the images presented here. We could not disagree more with some of the protesters, particularly those sympathetic to the US president, and yet we could not ignore them. Maybe the Internet itself has become a form of protest to all the preconceived ideas of class, public space, and employment?

E. Mattes · F. Mattes (✉)
318 Knickerbocker Ave, Apt 2H, Brooklyn, NY 11237, USA

© Springer Nature Singapore Pte Ltd. 2018
D. Cermak-Sassenrath (ed.), *Playful Disruption of Digital Media*,
Gaming Media and Social Effects, https://doi.org/10.1007/978-981-10-1891-6_9

Author Biography

Eva and Franco Mattes (1976) are an artist duo originally from Italy, working in New York.
Their medium is a combination of performance, video, and the Internet, for which they are perhaps best known. Their work explores ethical and moral issues when people interact at a distance, especially through social media, creating situations where it is difficult to distinguish reality from a simulation.

Melissa Gronlund, editor of *Afterall Magazine*, described Mattes' work as follows: "Whether by obscuring the name of the author, hiding information from the public, or presenting false information to (often unwitting) participants in the works they create, the Mattes set up situations in which the viewer's mistaken assumptions and actions create the form of the work itself."

Mattes' work has been exhibited at the Minneapolis Institute of Arts (2013); Site Santa Fe (2012); Sundance Film Festival (2012); PS1, New York (2009); Performa, New York (2009, 2007); AROS Aarhus Kunstmuseum (2009); National Art Museum of China, Beijing (2008); The New Museum, New York (2005); and Manifesta 4, Frankfurt (2002). In 2001, they were among the youngest artists ever included in the Venice Biennale.

They have given over a hundred lectures at universities, festivals, and museums, including Columbia University, New York; RISD, Providence; New York University; Carnegie Mellon, Pittsburgh; College Art Association, New York; Museo Reina Sofia, Madrid; MAXXI, Rome; and Musee d'Art Moderne, Paris.

They are founders and codirectors of the international festival, The Influencers, held annually at the CCCB, Barcelona, Spain (2004–ongoing).

The Mattes have received grants from the Walker Art Center, Minneapolis; The Museum of Contemporary Art, Roskilde; and ICC, Tokyo; and were awarded the New York Prize 2006 from the Italian Academy at Columbia University.

They were mentors of the New York Arts Practicum program (2012, 2014).

In 2015, they were faculty members at the MFA Fine Arts Department of the School of Visual Arts, New York.

Links to many of Mattes' projects can be found at their website: www.0100101110101101.org.

Part III
Mis-use, Struggle, Control

Sonification in an Artistic Context

Samuel Van Ransbeeck

Abstract After only 18 years, the 21st century might already be called the century of data. Internet usage has boomed and terabytes of data are being generated every day. Hence, it comes as no surprise that these data are used in ways other than originally intended. Data visualization as well as data sonification has become a widespread practice for scientific and artistic purposes. In the case of their use in artistic contexts, visualization and sonification are not limited to simply conveying information to users. Artists can use data to control specific elements of their works such as musical or visual parameters, reinterpreting the data, and creating awareness and engagement. In doing so, the artists transform abstract data into an aesthetic experience. This chapter focuses on sonification and discusses some projects that use the pair "data-sound" as a key building block. Looking at these projects, an aesthetic framework for sonification in an artistic context is then proposed.

1 Introduction

One can ask why a chapter on sonification is suited for this book. One of the reasons can be found in the source material that sonification uses. Stock values and temperature changes are all conveyed through numbers. These numbers get their significance from the meaning people attach to them. *For example: a temperature of 20° Celsius will be experienced as hot or cold, depending on the person, a distance of 100 kilometres will be experienced short for a plane ride but long for a walk.* Thus, numbers are contextualized, getting a meaning. The sonification artist takes values and attributes a new meaning to them. The original function of the data

S. Van Ransbeeck (✉)
CITAR | Centro de Investigação em Ciência e Tecnologia das Artes,
Portuguese Catholic University, Porto, Portugal
e-mail: thinksamuel@yahoo.com

© Springer Nature Singapore Pte Ltd. 2018
D. Cermak-Sassenrath (ed.), *Playful Disruption of Digital Media*,
Gaming Media and Social Effects, https://doi.org/10.1007/978-981-10-1891-6_10

is being subverted and turned around. A new meaning does not even have to refer to the original one, although there are compelling reasons to do so. If one takes the series of seismic shocks caused by an earthquake and creates an artwork referring to that earthquake, the work gets a double meaning: it functions to create awareness of a topic as well. For example, if one uses climate data, the artwork can make the audience aware of the climate change. Nevertheless, one can choose simply to use the data as control values for a work totally unrelated to the data's original attributed meaning. The artist's main goal then is to make an aesthetically compelling work.

2 Towards a Definition of Sonification Art

Sonification art requires the use of auditory displays. Therefore, prior to discussing or presenting sonification I contextualize it within the field of auditory displays (AD). This is followed by a presentation on algorithmic composition and in the end I propose a definition of sonification in an artistic context.

3 Auditory Display: Definitions, Types, and Functions

Sonification represents a subset of what we call an auditory display (AD). Similar to a visual display, an AD acts as an interface between an information source and a recipient. An AD is defined as "any display that uses sound to communicate information." This follows the same direction as de Campo (2008) who writes: "Auditory display is the rendering of data and/or information into sound designed for human listening."

These subsets of ADs can be distinguished:

1. The well-understood information subset. In auditory information displays, well-understood information is communicated through sound. Examples are public service announcements, auditory feedback sounds on computers, alarms and warning systems, and process monitoring systems. These messages do not require explanation: their meaning is immediately perceptible.

2. The sonification subset. Sonification or data sonification is the rendering of (typically scientific) data into (typically nonspeech) sound designed for human auditory perception. The informational value of the rendering is often unknown beforehand, particularly in data exploration. Through (repeated) listening to the sonic result the listener is able to grasp the meaning of the data.

Walker and Nees (2011) refine these distinctions in various ways according to their intended function, and distinguish several function categories in ADs (Fig. 1):

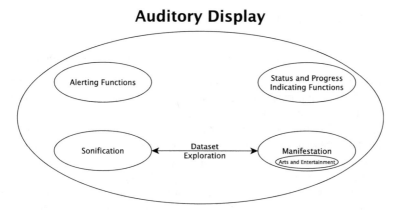

Fig. 1 Auditory display and its subtypes. *Sonification* and *manifestation* are connected through the dataset exploration function

1. Alerting functions: Any sound that warns the user of an event, for example, an alarm clock.
2. Status and progression-indicating functions: Sounds to keep the user informed about the status of a process.
3. Data exploration functions: Sound used to give a person an overview of the dataset as opposed to alerting and status-indicating functions that only show a momentary state.
4. Art and entertainment: In addition to yielding warnings and sonifications, events and datasets can be used as the basis for musical compositions. For example: one can generate musical sequences and build a composition with them.

Types three and four share similarities but have a different purpose. The distinction between data exploration functions (sonification) and sonification for art practices is also made by Polansky (2002) who divides the practice in a *Sonification* subset and a *Manifestation* subset; in *Sonification*, the objective is to sonify a process in order to better understand it. Its purpose is pedagogical and illustrative. An example is described by helicopter flight data being sonified and listened to in a compressed time space. Abnormalities in the flight data can be discovered by listening to changes in the sound. In *Manifestation* we are instantiating an aesthetic experience, in musical terms: a composition. Although alarm sounds or status report indicator sounds should not be seen as totally unaesthetic, the functionality, however, is primary in those types.

4 Sonification Modes

de Campo (2008) describes these modes of interaction with sonification:

1. Event-based sonification using parameter mapping: Each datapoint is mapped to a sound event. For example, each update of a stock price results in a new note where the frequency of the pitch is controlled by the price.
2. Model-based sonification: The user is able to interact with the data mappings. Instead of listening to a static representation of the data, he or she can manipulate parameters in real-time. The problem of interaction with auditory display is thoroughly discussed by Hermann and Hunt (2004).
3. Continuous sonification: In the case of high sampling rates, the data, such as seismic data (Bioacústiques 2010), can be translated into waveforms. This process is called audification. As the data patterns do not follow physical laws that exist in nature, the resulting waveforms can sound unearthly and are difficult to interpret for the user (Worrall 2009). On the other hand, these unearthly waveforms can be an interesting source for sound design and composition.

In sonification practice for data exploration, it is necessary to choose an appropriate mapping method. Listeners may experience difficulties in engaging with arbitrary mappings that have no relation to the data or cannot be easily related to the data. The difficulty in finding a suitable mapping is known as the *mapping problem*.

5 Sonification and Composition

Because manifestation can be related to musical composition, we focus on the exploration of the artistic possibilities of sonification in the context of ADs. In fact, we can correlate sonification (manifestation) with algorithmic composition, but we must highlight the differences.

Algorithmic music relies on an extra-musical process that is being sonified through mapping the data onto musical parameters. This practice is not unique to 20th and 21st century music. In the 14th century Guillaume Dufay used the proportions of the Santa Maria del Fiore cathedral in Florence to control the tempi in his motet *Nuper Rosarum Flores* (Taube 2004). In his *Musical Dice Game*, Mozart used two matrices of 8 by 11 cells. By throwing dice, a series of numbers is created. These numbers correspond to the 196 cells in the matrices, each containing a musical fragment. Each unique series thus results in a new musical piece. Although algorithmic, Mozart still adhered to the tonal doctrine. In the twentieth century, where traditional music paradigms were abandoned in favor of new experiments, composers such as Iannis Xenakis, Lejaren Hiller, and Gottfried Michael Koenig among many others started using algorithms in their music.

Many types of algorithmic music exist. Roads (1996) divides the field into stochastic and deterministic music. Maurer (1999) distinguishes between stochastic and rule-based music. In stochastic music, the composer uses a series of numbers generated by a computer. The composer still has quite some liberty in mapping the numbers to musical parameters. This is less the case in rule-based/deterministic music in which rules are established before the composition takes place. The best-known example of rule-based music is probable the treatise on counterpoint, *Gradus ad Parnassum* (Fux 1965).

Supper (2001) proposed a more refined distinction of rule-based music types:

1. The modeling of traditional compositional processes.
2. The modeling of new and original compositional processes, different from existing ones.
3. The use of algorithms through extra-musical processes.

Examples of the first type are described by McKay (2002) and Schottstaedt (1984). Ebcioglu (1990) created Choral, a program that harmonizes chorales in the style of Bach. *Illiac Suite*, created by Lejaren Hiller and Leonard Isaacson in 1957, is another example of the use of algorithms to model traditional compositional processes. A composition of the second type is *Çogluotobüsisletmesi*, by Clarence Barlow (1980). The metric and harmonic intensity, the uniformity in melody and rhythm, the density of the chords, attacks, and articulation were all calculated through algorithms. The resulting work is thus a composed meta-structure. Its final form is just one possibility of many final results (Supper 2001). The third type uses extra-musical processes to generate material, for example, Lindenmayer systems. Developed by Aristid Lindenmayer in 1968, a Lindenmayer system or L-system is a parallel rewriting system and a type of formal grammar. Lindenmayer used L-systems to describe the behavior of plant cells and to model the growth processes.

An L-system consists of an axiom and rewriting rules, for example:

Axiom: X, rule: X $\rightarrow$ XYX

If we apply this rule to the axiom, we eventually see this pattern emerge

X

XYX

XYXYXYX

XYXYXYXYXYXYX

XYXYXYXYXYXYXYXYXYXYXYXYXYXYXYX

An interesting characteristic of L-systems is the autosimilarity that occurs during the recursions of the system. Various composers have used L-systems, including Hazard et al. (1999) and Sodell and Soddell (2005).

What type of algorithmic music is *manifestation* (Polansky's term)? *Manifestation* takes an ever-changing input, which we cannot replicate when taking a new selection of data. Obviously, there are recurring patterns, but there are always subtle differences. It is impossible to encode these patterns and to set rules for their evolution. In that respect, we should consider manifestation a form of stochastic music. However, this is not entirely the case. Although we cannot be sure what the

Fig. 2 A third type of algorithmic composition

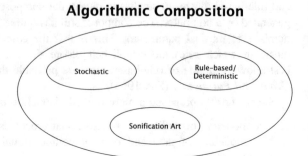

next datapoint will be, specific patterns occur, which make it different from random datasets. We are talking about an evolutionary system (type three in rule-based music) but we cannot encode this evolution; it is an ever-changing body of data. Especially in real-time musical works, it is difficult to create predetermined rules. A solution is to use interactive dynamic rules, which the user can manipulate in real-time. This helps her to adapt the work to new situations. We see the following problems that impede us from placing sonification sound art in either stochastic or deterministic music:

1. The use of real data as opposed to mathematical models.
2. Although it is difficult to predict, there is no randomness in the data. Patterns emerge but we can only see them after the data have been "created." There is a certain degree of information entropy.

As we can place this practice either in stochastic or deterministic music, I believe that we thus need to add a third category to algorithmic composition: sonification art (Fig. 2).

6 Towards a Definition of Sonification in an Artistic Context

Having compared separately the types of auditory display and algorithmic music, we can bring them together and position sonification art on the map. The AD types by Walker and Nees (see above) categorize arts and entertainment as applications of sonification. Algorithmic composition in itself is a subset of the arts, but not all algorithmic music is part of the sonification field (Fig. 3).

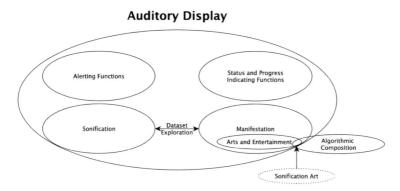

Fig. 3 The relation of algorithmic composition to sonification on the auditory display map. Sonification art is located in the common sector of arts and entertainment and algorithmic composition

7   Towards a Definition

In the preceding sections, I mainly discussed the differences between science and art as the goals or main approaches to data. One element, which is essential to artistic practice and remains to be discussed, is the appropriation by the artist of the data requiring creative and subjective aspects. As the artist uses data from observations of world phenomena and maps them onto musical parameters, it is possible that he will not be satisfied with the result. He can then decide to make alterations in the parts that are not to his liking. For example, in a series of values that are rising where one value is out of line, that drop can result in a broken melody, which might not be desirable in the composition. Thus, the composer may decide to alter the value to make the melody rise continuously. This results in a reinterpretation of the data: sonification as a creative trigger.

Creativity is an essential element in sonification. In my definition, I use Polansky's term "manifestation" to refer to the creative element in sonification. Xenakis is known to have done this type of local decision making, for example, in his piece *Herma* (Bayer 2000).

At the end of this section, we can finally propose a definition for sonification in an artistic context:

Sonification art is an arts practice that uses data from observations of world phenomena, which are mapped onto musical parameters through fixed or dynamic mapping procedures as an essential input element and that uses sound to manifest itself.

8 Data, Their Mapping, and the Sounding Result

A sonification process can roughly be divided into these stages:

1. Data fetching
2. Data mapping
3. Making the data sound.

When we talk about sonification, we first have to analyze our source material, namely the data. We can basically take any available dataset and sonify it. These data can come from various areas such as seismic data, weather data, or financial data. Many of these datasets can be found online, either for free (myriad datasets from the US federal government are available on http://www.data.gov) or for a fee (e.g., from the Microsoft Azure Data Marketplace). In a first stage we have to look for a dataset that is interesting enough to sonify. This means that we need a sufficiently dynamic system to study. Not all data are interesting; for example, the birth rate of the dodo in the last 200 years has been stable all that period. Thus it comes as no surprise that a sonification would not be very interesting. On the other hand, in a totally random number series, we would be confronted with such a sheer amount of data that it would be impossible to handle. Thus, it is paramount to choose an appropriate dataset.

The second stage is the way we are going to store and read those data. First we have to choose if we are going to use a live stream or a recorded dataset. Both have advantages and disadvantages. When using a live stream, we are confronted with the unexpected: we can never precisely know what will happen next. In his study of urbanism, Devisch (2008) calls such behavior *organized complexity*: through their various interactions, the constituent elements create a distinct macro-behavior, generating recognizable patterns or shapes (Johnson 2003). Hence, the mapping method needs to be sufficiently flexible to create an interesting sonification. This flexibility can happen on certain levels:

1. During the conceptual stage, where the artist tries out different mapping methods.
2. During the showing of the finalized artwork, where users can interact with the system, within the limitations set out in the conceptual stage.

When using a recorded dataset, we can read it several times and adapt the mapping method through multiple iterations. In real-time data, this repeated listening is impossible. However, both with historical and real-time data, the artist can decide whether to allow the user to interact with the choice of data and mapping methods.

Whether the work is interactive, and whether we use real-time data or a recorded dataset, the artist will try to make the artwork as compelling as possible, to the best of her judgment. If we want the user to listen for extended periods of time to the sound, we need to consider the aesthetic aspect of the sonification. Kramer (1994) writes that "Improved aesthetics will likely reduce display fatigue."

The third stage is making the data audible: how will the data sound? It is up to the composer to choose his instrument: he can choose to create a score to be performed by acoustic instruments or can map the data to synthesis parameters. Mazolla et al. (2011) say that the composition itself is not music but "a command structure for the production of sounding music." In this fourth stage the interpreting musician can make personal choices when reading the command structure (ibid.).

The four stages described above do not constitute a linear process: modifications to the processes in all stages can be made at any time. Hence, process and result are intertwined.

The goal of using data is eventually to interpret them, for instance, for scientific, artistic, commercial, or political purposes. The goal is to try to find the best or most compelling interpretation possible. Interpretation, however, does not equal understanding. In a scientific setting, the understanding of the data, the informational, is the most important aspect, whereas in an artistic rendering, the aesthetic experience is of paramount importance. Nevertheless, even in a scientific setting, aesthetics play an important role as they can have an influence on the listeners' fatigue.

9   Creative Processes in Sonification Art

The rise of conceptual art in the twentieth century has shifted focus from the artifact to the process of coming-into-being: how the artwork is created is at least as important as the resulting artifact. Danto (1998) proposes the term *posthistorical* to describe the period roughly from 1950 until now in which "art was no longer possible in terms of a progressive historical narrative" and exemplifies this with Andy Warhol's *Brillo Box*. Originally designed as a box to hold scouring pads, Warhol transformed the mundane object to an artwork. Indeed, for the *Brillo Box*, the conceptual part is more important than the piece itself. Warhol detached an object from its intended meaning and gave it a new meaning as an artwork.[1] In sonification art (or in algorithmic art in general), we detach the data from their original (informational) purposes and attribute new meanings to them. For example, stock market data serve to inform traders about the markets' movements; by sonifying them for use in an artwork, we detach the informational aspect from the data and we interpret the data differently to serve our artistic purpose. This does not imply that the artwork no longer refers to the informational aspect; the data are assigned a new significance.

How can we analyze the creative process? Mazzola et al. (2011) divide it into seven steps. The artist goes through these steps, constantly revising her work:

---

[1]Looking at the object was not sufficient to grasp its meaning: the audience had to know the concept behind it as well. This was clearly not the case when Canadian customs officials considered the work to be merchandise instead of an original sculpture and wanted to levy a customs duty when the artwork was imported into Canada (Danto 1998).

1. Exhibiting the open question: We start our quest by asking what we want to investigate.
2. Identifying the semiotic context: The question does not come alone. We need to find the surrounding context to be able to solve our question.
3. Finding the question's critical sign or concept in the semiotic context.
4. Identifying the concept's walls: What is limiting?
5. Opening the walls: Are the walls really necessary and how can we circumvent them to find new perspectives?
6. Displaying extended wall perspectives: Evaluating the perspectives that the artist found in the previous step and how these perspectives can be integrated in the work.
7. Evaluating the extended walls: Did the opening of the walls successfully extend the critical concept? If not, then the artist goes back in the chain and tries to find new walls and extensions or even a new critical concept.

We can divide the seven steps in two parts: questions 1–4, which deal with the initial conception of the work and questions 5–7, which extend the work's possibilities. We have to ask now if we can apply this way of questioning to sonification art. But first, I present several artworks to contextualize the practice of sonification art.

10   Some Example Works

In this section, I discuss a selection of artworks that use data in different ways from which they were intended, effectively subverting their original purpose.

10.1   Projects Using Environmental Data

One category of projects uses data measuring natural phenomena such as meteorological data, astronomic data, seismic data, and the like. Andrea Polli's *Atmospherics/Weather Works* (2004) is an installation that uses meteorological data from a snowstorm on February 18, 1979 on the US east coast. Data collected in 16 locations at five different elevation levels are used and assigned to a 16-channel sound system. The data sampling rate was one datapoint every three minutes during 24 h. The variables—atmospheric pressure, water vapor, relative humidity, dew point, temperature at each elevation level, and total combined wind speed of all five levels—are then mapped onto musical parameters.

Quinn and Meeker (2001) created *The Climate Symphony*, a performance/presentation. This work uses information about the environment from an ice core drilled up in Greenland containing information from the past 110,000 years. Solar intensity, ice sheet movements, and volcanic activity are mapped onto patterns played on vibraphone, tom-toms, cowbells, crash cymbals, and timpani. The earth's wobble (changes in the earth's axis) is mapped onto an organ sound. The tilt of the earth is mapped onto a three-note arpeggio whereas the oceans' 1450 year cycles (periods of relative climatic stability) determine the instrument.

The Radioactive Orchestra (Livet 2011) is a system that sonifies radioactive energy coming from gamma rays that are emitted with decaying radioactivity levels. Mapping is straightforward: the higher the energy level, the higher the pitch.

Roden and Polsenberg's (2004) Ear(th) installation uses data from a 1999 earthquake in the Mojave desert. A series of glockenspiels are located in a space and each glockenspiel's note is played according to the value read. A prerecorded version of the piece is played together with the real-time playing installation to create a polyphonic effect (Roden 2012; Thomas 2012).

10.2 Projects Using Social Network Data

Harman's *Twinthesis* (2011) uses the ASCII characters of public tweets to control sound. The values of the characters are sent to an oscillator bank and mixed with a random value to create a wide variety of frequencies. Furthermore, the tweet properties are used to control granular synthesis parameters. The length of the tweet (with an added random value of +5/–5) controls the buffer size of the sample.

The *Sound of Mercadolibre* (Nussbaumer and Kachel 2007) was programmed in Supercollider. Real-time data are read from Mercado Libre, a South American online trading platform similar to eBay. Notes are chosen from 11 possible scales in a range of three octaves. Sonified parameters are the feedback score, the number of feedbacks, the item being traded, and the user type (seller or buyer). In addition, 20 comments from the feedback are selected at random and are read by the MacOS speech synthesizer. A nonreal-time version, using data from eBay, was created as well.

Rhythms of the City (Guljajeva and Sola 2011) is an installation that analyzes geodata obtained from YouTube, Twitter, and Facebook. The measured activity in a city controls the speed of a metronome: the more activity, the higher the frequency. The installation consists of an array of 10 metronomes, each linked to a different city. This results in a polyrhythmic soundscape.

10.3 Projects Using Financial Data

Financial data are one of the most dynamic types of data that are created by humans. Every second, transactions of billions of Euros and other currencies are being conducted.

StockWatch (Van Ransbeeck and Guedes 2009) takes stock market indices (e.g., the FTSE100 and Nasdaq100) as its input. The value of a stock is mapped onto a

MIDI-note. Because an index is used as input, a sequence of stocks passes and the note sequence varies infinitely (if the stock price changed after the previous appearance). StockWatch has evolved into *DataScapR* (Van Ransbeeck 2015), a toolbox of MaxMSP patches to use stock market data in musical creation. The user can sonify near-real-time data as well as historical data. Furthermore, using the Bach objects (Agostini and Ghisi 2012), one can create symbolic scores.

In *Stock Exchange Piece*, Mathieu Saladin (2007) took the daily price and the 50 day moving average (the average price over a period of the last 50 days) of crude oil and gold during 50 days (between the 4th of March and 22nd of April 2007, with a resolution of one datapoint for a day) and mapped them one-to-one (or unit-swapping, as Saladin calls it) onto sinewaves. The prices are panned to the left and the 50 day MA sinewaves are panned to the right. Each day lasts 1 min, which allows the listener to experience the acoustic phenomena resulting from the slow-moving sinewaves. These movements make the composition into a 50 min slowly evolving, minimalist drone piece.

The Spanish group Derivart (2008), whose works are focused on socioeconomic topics, developed *Spreadplayer*, an interactive table where the user can hear stocks from the Madrid stock exchange by dragging them into a radar like space.

11 Towards an Aesthetic Framework for Sonification Art

The works described above use sound as a key element to convey an aesthetic experience. Contrary to scientific sonification, the primary goal is not to make the data comprehensible. This aspect can even be totally absent in artistic projects. In this regard Roden and Polsenberg (2004) says of his installation work *Ear(th)*:

Just as the installation would look and sound different if the data source was the ocean or Shakespeare; Ear(th) in no way attempts to illustrate the earth's movement or to recreate an earthquake experience through sound. I am much more interested in simply allowing the earthquake data to generate a sound composition and to allow for my own misreadings of the data to suggest placements, sound ideas, performances, and sculptural forms. For me, this process is a kind of alchemy—to allow the materials to be transformed into something completely connected to, yet seemingly distant from, the source.

The artwork attaches a specific meaning to the data; it superimposes a new additional meaning over the data's original meaning. Thus, the artist repurposes the data's original goal: instead of making the development of a stock's price understandable, the data value itself is used to give significance to a musical parameter. Working with an artistic goal in mind, the artist does not have to make her sonification neutral: unlike in scientific sonification, in sonification art, the focus lies on the underlying message or aesthetic experience the artist wishes to convey.

Earlier in this text, I discussed Mazzola et al.'s seven steps and divided them in two parts. I apply these questioning parts now to sonification art.

In exhibiting the question, we ask ourselves if a certain dataset could be used to create compelling music. In order to do that, one needs to discover a method to translate the data into sound. Here, the question's critical concept relies upon the understanding that we use data as the starting point of the work.

Following the concept and the context within which our concept is located, the next step is to find the limitations of the concept. It is possible that the dataset is not interesting or the mapping methods do not result in compelling music. The artist may then either change the dataset or transform the musical output (by using different mappings). For example, if the artist wants to use seismic data, probably a dataset with lots of seismic activity will be more interesting to explore than a relatively calm seismic period. By changing the dataset, the artist can thus find different outputs. If the artist is satisfied with the dataset, there is the possibility of transforming the raw output. For example, the artist can try out different mappings and make a collage of the interesting parts. By going through a system of incremental revisions of the work, the artist gains more insight into his practice and can come to the desired realization of the artifact.

Sonification art often happens through incremental revisions: it is obvious that the work does not consist solely of the final product but involves a particular system, designed to create the artifact. Composition becomes a form of *system design* with the musical artifact being the traces of that design. If we apply this to a sonification work we can see different stages where the modifications can be performed:

(1) The data themselves: The choice of data is quite important. The artist can choose to change her dataset if (for example) the data are not varied enough.
(2) The mapping of the data: One can experiment with different mappings and choose the most compelling mapping method to her liking.
(3) The sounding artifact: After having designed the complete system, the sounding artifact can still not be satisfying to the artist. She can choose to either leave it as it is and consider the bad-sounding artifact as a natural result of the process, or she can make local changes (she can, e.g., change a few notes if that sounds better).

Each of these revisions implies a phase in the incremental creation of a desired artifact. This process does not have to happen in a linear fashion: we can change the mapping, decide later that the data are not interesting, and work on the mapping again with a new dataset.

Using extra-musical material can result in finding new ways of musical expression. When using extra-musical data, we are confronted with a sequence that will not automatically result in a logical (traditional) musical progression (be that melodic, rhythmic, or another musical parameter) when mapped to music. This does not mean that the resulting music will be bad, quite to the contrary. Using sonification techniques one can discover new musical progressions. Doornbusch (2005) says of algorithmic composition (and this can be applied to sonification as well): "The work with computers in this way has caused me to have compositional ideas that I would not otherwise have had." These new ideas will lead to new compositions, conveying a new way of artistic creation. As such, we can say that sonification is a way of inventing new music through the repurposing of data.

12 Conclusion

In this chapter, I discussed sonification art: I proposed a definition of sonification art and placed it within the realm of algorithmic composition. I then described creative practices and presented some artworks to give an overview of the practice. I applied the conceptual elements of creativity to sonification art and, finally, I discussed a few aesthetic considerations to provide a foundation for an aesthetic framework for sonification art.

Acknowledgements I would like to thank my supervisor António de Sousa Dias for his continuous (incremental!) revisions and careful guidance in writing this chapter. This research is sponsored through a Ph.D. scholarship by the Portuguese Foundation for Science and Art with reference SFRH/BD/72601/2010.

References

Agostini, A., & Ghisi, D. (2012). *Bach: An Environment for Computer-Aided Composition in Max* (pp. 1–6).

Barlow, C. (1980). *Bus journey to Parametron: All about Cogluotobusisletmesi* (Vol. Feedback papers). Cologne: Feedback Studio-Verlag.

Bayer, F. (2000). *De Schönberg à Cage.* Klincksieck.

Bioacústiques, L. D. (n.d.). LIDO: Listening to the Deep Ocean Environment. Retrieved November 20, 2010, from http://listentothedeep.com/.

Danto, A. C. (1998). Beyond the Brillo box: The visual arts in post-historical perspective (3rd ed.). University of California Press.

de Campo, A. (2008). Toward a data sonification design space map. In G. P. Scavone (Ed.), *Presented at the Proceedings of the 13th International Conference on Auditory Display* (pp. 343–347). Montréal.

Derivart. (2008, May 30). Spread.player. Retrieved October 2, 2013, from http://www.derivart.es/index.php?s=p7&lang=en.

Devisch, O. (2008). Should planners start playing computer games? arguments from simcity and second life. *Planning Theory and Practice, 9*(2), 209–226. https://doi.org/10.1080/14649350802042231.

Doornbusch, P. (2005). Pre-composition and algorithmic composition: Reflections on disappearing lines in the sand. *Context: Journal of Music Research, 47*(29/30).

Ebcioglu, K. (1990). An expert system for harmonizing chorales in the style of JS Bach. *The Journal of Logic Programming, 8*(1), 145–185.

Fux, J. J. (1965). *The Study of Counterpoint from Johann Joseph Fux's Gradus Ad Parnassum.* (A. Mann & J. Edmunds, Trans.). New York: W. W. Norton & Company.

Guljajeva, V., & Sola, M. C. (2011). The Rhythm of City. Geo-located social data as an artistic medium. In *Presented at the International Symposium for the Electronic Arts* (pp. 1–5). Istanbul.

Harman, S. (2011). *Twitter Powered Synthesis. B.Sc. Degree in Music Audio Technology* (pp. 1–11).

Hazard, C., Kimport, C., & Johnson, D. (1999, October 1). *Fractal Music.* Retrieved October 1, 2013, from http://www.tursiops.cc/fm/.

Hermann, T., & Hunt, A. (2004). The importance of interaction in sonification, (pp. 1–8). Retrieved from http://hdl.handle.net/1853/50923.

Johnson, S. (2003). *Emergence: The connected lives of ants, brains, cities, and software.* New York, United States: Scribner.

Kramer, G. (1994). An introduction to auditory display. In G. Kramer (Ed.), *Studies in the Sciences of Complexity Proceedings* (Vol. 18, pp. 1–78). Readin, MA: Santa Fe Institute.

Livet, K. (2011, October 2). *Radioactive Orchestra.* Retrieved October 2, 2013, from http://www.radioactiveorchestra.com/.

Maurer, J. A. (1999, March 1). The history of algorithmic composition. Retrieved October 1, 2013, from https://ccrma.stanford.edu/~blackrse/algorithm.html.

Mazzola, G., Park, J., & Thalmann, F. (2011). *Musical Creativity.* Berlin, Heidelberg: Springer.

McKay, C. (2002). *Final Project Report SpeciesChecker 1.0.* Montréal: McGill.

Nussbaumer, S., & Kachel, E. (2007, November 28). Sound Of Mercadolibre—Ubermorgen.Com —Text To Sound Generation—Robots Chaos Structure—Mexico. Retrieved October 2, 2013, from http://www.sound-of-mercadolibre.com/.

Polansky, L. (2002, April 15). Manifestation and sonification. Retrieved September 1, 2013, from http://eamusic.dartmouth.edu/~larry/sonification.html.

Polli, A. (2004). Atmospherics/weatherworks: a multi-channel storm sonification project. In *Presented at the Proceedings of the 2004 International Conference on Auditory Display.* Sydney.

Quinn, M., & Meeker, L. D. (2001). Research set to music: The climate symphony and other sonifications of ice core, radar, DNA, seismic and solar wind data. In *Presented at the Proceedings of the 2001 International Conference on Auditory Display* (pp. 56–61), Espoo.

Roden, S. (2012, January 20). Question about Earth.

Roden, S., & Polsenberg, A. (2004). Ear(th). Retrieved from http://ddata.over-blog.com/xxxyyy/1/96/04/42/Textes-1/steve-roden-earth-installation-at-art-center.pdf.

Saladin, M. (2007). *Stock Exchange Piece.* (Mattin, Ed.) *mattin.org* w.m.o/r (32nd ed., Vol. 50).

Sodell, J., & Soddell, F. (2005, December 28). Microbes, L-systems and music. Retrieved October 1, 2013, from http://cajid.com/jacques/lsys/index.htm.

Schottstaedt, B. (1984). Automatic species counterpoint (No. STAN-M-19). *CCRMA reports* (p. 70). Stanford. Retrieved from https://ccrma.stanford.edu/files/papers/stanm19.pdf.

Supper, M. (2001). *A Few Remarks on Algorithmic Composition, 25*(1), 48–53. https://doi.org/10.1162/014892601300126106.

Taube, H. (2004). *Notes from the Metalevel: An Introduction to Computer Composition.* London, UK: Routledge.

Thomas, A. M. P. (2012, January 20). Question about the Earth work with Steve Roden.

Van Ransbeeck, S. (2015). Transforming the Stock Markets into Music using DataScapR, *7*(4), 1–12. Retrieved from http://piim.newschool.edu/journal/issues/2015/04/index.php.

Van Ransbeeck, S., & Guedes, C. (2009). Stockwatch: A tool for composition with complex data. *Parsons Journal for Information Mapping*, 1–3.

Walker, B. N., & Nees, M. A. (2011). Chapter 2 Theory of Sonification. In T. Hermann, A. Hunt, & J. G. Neuhoff (Eds.), *The Sonification Handbook* (pp. 9–39). Berlin, Germany: Logos Verlag.

Worrall, D. (2009). Sonification and information. In *Sonification and Information: Concepts, Instruments and Techniques.* Canberra, Australia.

Author Biography

Samuel Van Ransbeeck is a Belgian composer and researcher. He works mainly with algorithmic compositional approaches and focuses on interdisciplinary art, collaborating with visual artists. He got his PhD from the Catholic University of Porto, Portugal, where he developed *DataScapR*, a toolbox for sonification in an artistic context. His main research interests are sonification, algorithmic composition and generative music systems.

Little Big Learning: Subversive Play/GBL Rebooted

Chad Habel and Andrew Hope

Abstract Game-based learning is a buzzword heard with increasing frequency in educational technology circles, but these discussions often proceed with an insufficient understanding of the nature of play in a social and cultural context. This chapter problematises some common approaches to game-based learning by exploring social dynamics and relations of power to propose a more critically disruptive model of game-based learning. Using the *Little Big Planet* franchise as a case study, it argues that game-based learning serves little purpose if it replicates authority-centred, transmissive ideas of learning, and that focussing on players/students as the producers (not just consumers) of digital texts for learning is significantly more productive.

1 Introduction

Game-based learning is becoming considered as a serious alternative to traditional classroom instruction, evident in its presence in the two-to-three year adoption horizon of the NMC Horizon Report (Johnson et al. 2013). However, educators may not be ready for the challenges to their own practice and conceptions of learning that are implied by such a shift.

In considering game-based learning (GBL), some see the lurking danger of trivialising learning, but the notion of serious games suggests that such a threat can be avoided. As early as 1982, Ken Jones wrote about using simulations to assess student activity, rather than continuing to use the students to assess the simulations. In particular, he railed against 'limited research' and the 'rose-tinted glasses' of some researchers (Jones 1998: 331). More recently, Charsky has highlighted the

C. Habel (✉)
Teaching Innovation Unit, University of South Australia, Adelaide, SA 5000, Australia
e-mail: chad.habel@unisa.edu.au

A. Hope
School of Social Sciences, University of Adelaide, Adelaide, SA 5005, Australia
e-mail: andrew.hope@adelaide.edu.au

© Springer Nature Singapore Pte Ltd. 2018　　　　　　　　　　　　　　　167
D. Cermak-Sassenrath (ed.), *Playful Disruption of Digital Media*,
Gaming Media and Social Effects, https://doi.org/10.1007/978-981-10-1891-6_11

danger of seeing games as 'motivational holy grails' (2010: 193), whilst recognising that many zealous proponents of new GBL methods do more harm than good. Indeed, it is essential to avoid repeating the mistakes of 'edutainment', which embodies the 'worst type of education' (Charsky 2010: 177). As Squire and Jenkins put it, '[f]rankly, most existing edutainment products combine the entertainment value of a bad lecture with the educational value of a bad game' (2003: 8). The problem is that edutainment mostly consists of 'drill and practice', or 'surface' rote learning activities, which are widely considered to occupy the lowest rung on taxonomies of cognitive activity (Krathwohl 2002; Biggs 1999). It is the digital equivalent of reciting times tables, which admittedly is better than nothing, but is still extremely limited.

Nonetheless, well-designed commercial games have the potential to provide valuable learning experiences, as long as they are designed effectively (Whitton 2010). For instance, Pac-Man 'is an action system where skills and challenges are progressively balanced, goals are clear, feedback is immediate and unambiguous and relevant stimuli can be differentiated from irrelevant stimuli. Together, this combination contributes to the formation of a flow experience' (Bowman 1982: 15, cited in Squire n.d.). Consequently, a widespread conception of game-based learning involves the use of predesigned games in a formal learning environment, often with some scaffolding from a teacher or educator. An array of educational theories around curriculum and assessment design can help to explore the potential of games used in this way. For example, games provide an environment in which formative assessment is deeply built into the activities that students/players undertake, and are therefore able to set up instant feedback loops to shape learning. Examples of the highly popular commercial game franchise *Little Big Planet* demonstrate the potential for predesigned virtual game environments to facilitate learning in a way that is fun and engaging. Nevertheless, this approach only hints at the potential of game-based learning, which can be significantly enhanced with a more critical approach. A key problem with this traditional notion of game-based learning is that it still relies on an expert educator/game designer to create a resource that students use in a relatively passive manner, which differs little from traditional 'teacher led' education techniques. Anthropological notions of learning as play help to explain how game-based learning can subvert the power relations inherent in traditional instructional environments. Ultimately, it is not *games* as such, but *game platforms* that have the true potential for transformative learning. In this sense, the level editor function of *Little Big Planet* may show more potential for learning than even the best-designed learning levels.

2   *Little Big Planet* as a Learning Space

Comprising of *Little Big Planet* (2008), *Little Big Planet 2* (2011) and *Little Big Planet 3* (2014), *LBP* is a highly successful commercial game franchise that has led the way in providing accessible user creation tools. At heart, it is an accessible platform game in the tradition of *Super Mario Bros.* The player's avatar, or 'SackPerson', is required to traverse a variety of hazards, dispatch opponents and solve puzzles in order to complete set missions. Admittedly, some might not see the value of using a game like *LBP* for the lofty goal of uniting gaming and learning (Shute et al. n.d.). Indeed, some could perceive *Little Big Planet* as incapable of rising about the 'perceived triviality' associated with some educational games, and commercial games in particular (Woods 2004: 14). Furthermore, the attempt to imbue *LBP* with learning might be seen as 'sugar-coating' learning, or producing 'mere recreations of commercial games with learning inserted' (Charsky 2010: 193). This perspective could label computer games as having an unsupportable focus on entertainment or fun, at the expense of learning.

Nevertheless, this criticism rests on a false dilemma between learning and fun, which pervades even the most scholarly and well-intentioned writings on educational gaming. For instance, Shute et al. (2011) define serious games as being designed to educate not entertain—this assumes that the two are, or can be, mutually exclusive. In fact, the best learning, even in traditional forms, is fun—it is at once entertaining and educational. We should also be wary about Charsky's (2010) predilection for seeing 'commercial' games as a type of contamination that may dilute 'pure' learning. Gee (2007) demonstrated most eloquently that even 'commercial' games can provide powerful insights into learning and literacy, and Whitton (2010) argues that commercial games have particular affordances that make them highly valuable as learning applications.

Before considering the links between *LBP*, game play mechanics and rewards/feedback in more detail, it is vital to consider the notion of play and how it has been historically linked to learning. This will establish a basis for the argument that students should be encouraged to playfully design their own content as part of a learning experience which is more engaging than simply playing predesigned levels.

3   The Concept of Play

Play is a problematic notion to define. As Sutton-Smith (1997: 1) notes '[w]e all play occasionally, and we all know what playing feels like. But when it comes to making theoretical statements about what play is, we fall into silliness. There is little agreement among us, and much ambiguity'. Partly this reflects the fact that the notion of play is socially and culturally constructed, arising through lived experiences. Links between play and learning within educational institutions are inherently political in nature. As Rieber (1996: 45) suggests, views relating to the acceptability of play in the educational process can be fluid: '[a]s the prevailing philosophy in education changes, so too does the attitude towards play. In one era, play can be viewed as a productive and natural means of engaging children in problem-solving and knowledge construction, but in another era it can be viewed as wasteful diversion from a child's studies'. The current contentious position of GBL in formal curricula and educational institutions may be a reflection of such political whims.

Nevertheless, certain distinguishing features of play in formal learning processes have been proposed. Pellegrini (1995) suggests that play is usually voluntary, is enjoyable in its own right, is not dependant on external rewards, includes active engagement and has a make-believe quality. Yet such qualities have perhaps encouraged individuals to regard play and work as opposites. As Blanchard and Cheska (1985) suggest this is misleading, insofar as work can be playful, particularly if job satisfaction is so high that monetary reward is a secondary concern. This suggests that learning as part of a work process can upon occasion also be play, so a key issue might be how to engender play in work situations. Alternatively, as play is a central feature of computer gaming, an alternative approach might be to consider how to reinforce learning elements and introduce a component of work in such entertainment media.

A more useful way to approach the concept of play might be through considering four key themes that have emerged from contemporary theories on this subject. Namely, play as progress, play as power, play as fantasy and play as self (Sutton-Smith 1995). Play as progress suggests a cognitive or educational purpose, wherein the individual learns something 'useful'. Deciding what knowledge should be labelled as 'useful' is at the very heart of debate about the efficacy of GBL, with traditionalists narrowing notions of learning and rejecting commercial games as educational spaces. Play as progress is often linked to the process by which children 'become' adults, creating a false dichotomy between the play of children and adults. Furthermore, such a view adopts a deficit model of childhood, stressing potential outcomes, whilst dis-privileging the agency of children. Play as power suggests competitions within which losers and winners are declared. Whilst this is often a key element of computer game playing, a broader interpretation of the notion of power allows for a deeper consideration of issues around GBL, such as the role of commercial companies, established educational institutions or the extent to which children have a voice in expressing how they would prefer to learn effectively. Play as fantasy suggests creative and imaginative thinking, with obvious links to play as progress. As this element is perhaps most strongly linked with image it could also have connections with social identity, which is often creatively negotiated within and between people's life worlds. Finally play as self emphasises intrinsic worth, the quality of experience. This could be interpreted as 'entertainment value' and consequently is a key element of a well-designed computer game, but could also involve 'shifting between' multiple selves (see Penney and Mueller in this volume).

These four themes all offer insights into the social dynamics around educational processes in GBL, as well as the role of play in learning, engagement and identity formation. Yet play has much greater potential to impact positively on the learning process than that already suggested. Before further exploring the issue of play and user-generated content, it is necessary to make a stronger connection between the notion of play, computer games and learning. Gee (2007) suggests computer games can teach us about effective learning strategies, and how people learn. According to Shute (2008) a core mechanism is formative feedback, which can be seen as an essential part of any learning endeavour. Considering feedback, and its formalised embodiment in the form of assessment, will not only provide insights into how commercial games can be constructively appropriated and used as learning spaces, but will also foreground some issues that need to be considered when the focus of this paper shifts towards user-generated content.

4 Game-Based Learning, Rewards and Feedback

Feedback is a key element in seeking connections between learning spaces. In simple terms, feedback refers to the transmission of evaluative information about an action or process to the original source. Whilst in formal educational spaces feedback that takes the form of assessment is often privileged, in computer games a focus on rewards as formative feedback is much more common. For example, in the *LBP* games there are score bubbles that represent a reward proportional to the task, puzzle or technique the player has completed. Significantly, score bubbles are often used as instructional cues, particularly in tutorial levels to learn the basic mechanics of the game. Yet, score bubbles are more than just feedback. They are also used to create competition between players in multiplayer modes and for measuring performance in such competitive contexts. Game trophies attached to a user profile serve the same function. Such rewards are a type of feedback which is often crucial in the learning process, adding entertainment value but also providing comparative metrics. It is also possible to further distinguish between diagnostic (before learning), formative (during learning) and summative (after learning) elements (Crisp 2007). These considerations of rewards, feedback and assessment are significant if we consider play as progress, as feedback can be seen as an essential element that scaffolds learning through play.

Some applications of game-based learning have led to an innovative notion of assessment that has not previously been considered in educational theory, and could arguably only be possible in online learning and educational games. This is the idea of 'embedded formative assessment' or 'stealth assessment' (Shute et al. n.d.), where the assessment of student activity takes place within the learning system and largely without the awareness or knowledge of the student. In this type of assessment design, 'the assessment cycle continues in very small loops that provide immediate, detailed and unobtrusive formative feedback. In this way, the assessment structure is ubiquitously embedded in the system, and the examinee/player does not experience interruptions to instruction for separate assessment activities' (Behrens et al. 2007: 66). Stealth assessment, as a kind of ideal formative assessment, provides key information to the system regarding the student/player's activities and achievements in relation to the learning objectives. Because it is seamless and unobtrusive, it is particularly designed to induce (and retain) 'flow', or the player's immersion in the gaming/learning environment (Csikszentmihalyi and Csikszentmihalyi 1990). This notion of flow links directly with the vast, and growing, literature on engagement in education (see especially Whitton 2010).

Whilst the links between stealth assessment, flow and engagement may make sense to many educators, it must be said that the principles behind stealth assessment are contrary to much current educational theory and practice. Current best practice dictates assessment structures and criteria must be explicitly stated. Educational institutions require curriculum detail, and criteria-referenced assessment design requires that individual pieces of assessment be designed with listed criteria (i.e. rubrics) which are not only used by assessors for marking work, but also communicate expectations about assessment to students. Sometimes they are also used in self- or peer-assessment activities. The reasoning behind this is that if students clearly understand what is required of them, they are in a strong position to effectively fulfil those requirements. Therefore, the notion of stealth assessment may seem unusual (or even somewhat alarming) to educators. Nevertheless, stealth assessment operates in *LBP*. The learning of game mechanics (including the 'create' mode) operates without the player's conscious awareness, and information is covertly fed back into the system to further shape the learning environment in order to support the player's learning.

Drawing on notions of feedback and curriculum design, Shute et al. (2011) have demonstrated how evidence-centred design (ECD), 'a framework for developing assessments by systematically linking models of understanding, observable actions and evaluation rubrics to provide evidence of learning' (Shaffer et al. 2009), can be used to identify valued competencies such as creative problem-solving in games as diverse as CISCO network training and *The Elder Scrolls: Oblivion*. From an ECD perspective, rewards are fed directly back into the system to alter the learning environment to help the student learn (formative assessment) and provide the teacher/game/instructional system with accurate information on the final performance of the student (summative assessment). ECD allows a more detailed focus on assessment as a crucial learning mechanism in serious games, and even the design of commercially successful computer games can be elucidated using its concepts. For example, the earlier level designs of *LBP* revolve closely around learning objectives, which are made explicit to the player/learner: they may include simple controls such as moving, jumping and grabbing objects and progressively move up towards more advanced functions such as riding transport (a skateboard or wooden horse) and using sticker triggers. These activities are precisely aligned with the objectives of the levels. Upon completion of these activities, the player is rewarded in a variety of ways: with points, objects and progression to future levels. Therefore, the learning environment is integrated and assessment data (rewards) are fed directly back into the system to condition progression.

Ultimately, feedback is more than merely an indication of play as progress. Drawing upon broader notions of play as fantasy, it can be seen that certain types of rewards can also potentially lead to identity formation. Ownership mechanics are applied in *LBP* by providing each player the ownership of a game pod (essentially a menu space), a 'SackPerson' avatar and a level builder for them to personalise and make their own. Costume rewards are used by players to dress and personalise their 'SackPerson', creating a 'projected identity' of the player through an avatar (Gee 2007). Furthermore, in-game reward 'stickers' offer players a greater range of colours and materials to 'paint' and decorate their gaming pod (home screen), custom levels and even their avatar. This provides an avenue to express one's self through customising the game environment. Stickers and costume prize bubbles rewarded for playing the game are offered to the players as creative material that allows control through design of their gaming environment. An anthropological approach would see this as 'play as self' (Sutton-Smith 1995), whilst Gee (2007) would identify this as 'projective identity'.

Even more significant from a GBL approach are the rewards found in *LBP* that allow players not only to customise, but also to create new game levels. Objects are awarded to players throughout the game that provide templates and/or building materials to assist with creating levels in the level builder. Building and sharing a level is a large motivator to 'playing' *LBP* as it offers the player a blank canvas to express themselves in a gaming context. Therefore, objects become desirable rewards for players who want to contribute and express themselves in *LBP*. The strategic use of costumes, stickers and object rewards in *LBP*, together with its strong community of players, has generated millions of user-created levels, personalised costumes to customise avatars and most importantly provided control (ownership) of the game in a creative way to all players. It is this control nexus that provides the subversive potential of learning through games, (re)appropriating the educational process, and rebooting GBL as we currently know it.

5 *LBP*: From Content Delivery to Digital Construction

On a basic level, *LBP* provides the opportunity for players to complete levels, which teach them something through play. Crucially, this learning through play is heavily mediated by the process of design and particularly crafting, deeply embedded in the act of play (see Nitsche in this volume). Therefore, even more important than the accessible gameplay is the accessible level design. There are two ways players may learn through *LBP*: by playing previously built levels designed to instruct and by actually building levels themselves. Media Molecule's purpose was to design a game which would allow user-created content to be generated by anyone and everyone: no particular game design skills would be necessary, nor any special technical knowledge.[1] This puts the power of creating learning environments into everyone's hands, not just an expert educator's. A consequence of this accessible level design is the development of user-creator communities. These may be considered 'communities of practice', a term used to describe 'situated learning' or learning which occurs in a specific social and cultural context (Gee 2007; Lave and Wenger 1991). This is not something that is tangential to effective learning

---

[1]It should be noted that these platforms for content generation are becoming more and more popular: witness *Scratch*, *Greenfoot*, *Minecraft*, *Disney Infinity*'s toybox mode and Microsoft's *Project Spark*.

through games; in fact, it is central to it. Time and again authors emphasise the instructional context of learning by gaming (Arnseth 2006); that research into these instructional contexts is very much needed (Squire 2002); and that '[l]ike reading and thinking, learning is not general, but specific; like reading and thinking, it is not just an individual act but a social one' (Gee 2007: 7). *LBP* therefore allows for this communal, social and cultural construction of learning through the community of practice of user-creators on the PlayStation Network.

It is now possible to explore some of the user-created content of *LBP* to see how effectively these levels support learning. A quick search of the *LBP* community levels finds many levels that aim to teach players a variety of skills and abilities, from simple ABCs to more advanced grammar to different languages, from basic arithmetic to more advanced algebra, from how to use logic design elements in *LBP* levels to the use of Boolean logic. Indeed, whole suites of levels have been designed to support learning in STEM and STEAM (Grant 2011; ConnectED 2013).

A brief review of some of these professionally designed learning levels gives an idea of the kind of experience they offer. The *Discovery Pier* suite of levels, crafted as a theme park, aims to instruct users on aspects of classical mechanics such as gravity, g-forces and rotational motion, though explaining the concepts in texts boxes as they relate to rides that the player can then go on and see the principles in action. The *Stem Cell Sackboy* level explains basic human biology and especially the function of stem cells in human development: this is done through demonstration and some interactivity. Most impressively, *Aeon Quest: Abduction* sets a series of challenges around arithmetic, geometry and logic puzzles and provides a kind of instruction, opportunity to practice, even and rudimentary formative assessment.

No doubt, these levels are well designed (especially *Aeon Quest*) but ultimately they work within a didactic pedagogical model. They are essentially a resource (albeit highly interactive) designed by an expert for the consumption of a student. The overarching dependence on the articulation of knowledge through an avatar, especially as it almost always happens in text boxes, makes most of the learning very much like a lecture or textbook, delivering knowledge which, perhaps, is then applied to challenges later. This is hardly the brave new world of learning which is sometimes hailed by GBL pundits.

Whilst this user-generated content illustrates how commercial computer games can be used to promote learning through play as progress, they are created with a specific, limited learning task in mind. Furthermore, despite the appeal of play as progress, buttressed through the various awards on offer, there is the danger that such content might repeat some of the failings of edutainment games. Learning through play need not be simply seen through the lens of play as progress: in fact, if we see play as power, then the potential for GBL to fundamentally challenge institutional power structures becomes more apparent. Nonetheless, we may even move beyond this somewhat limited notion of play as power.

6   Dissembling Play and the Creation of Learning Spaces

Whilst play can be described in terms of progress, power, fantasy or self, it should also be recognised that it is a 'free activity' through which individuals can learn and improvisation can be inspired (Schechner 1994: 621). Significantly, the learning potential of play stretches beyond the formal curriculum. Play can be inventive, challenging social boundaries, whilst blurring distinctions of reality/unreality (Ingold 1994). Consequently playful learning activities can test and confront existing social norms, including the power relations embedded in formal learning environments. As Turner argues '[p]layfulness is a volatile sometimes explosive essence' (Turner 1983: 233).

From this perspective it can be argued that rather than advocating that 'educators' build *Little Big Planet* user-generated content so that 'learners' can then be forced through somewhat contrived, learning-related scenarios, students should be encouraged to create their own worlds. Dissembling boundaries and engaging in playful possibilities, world creation allows students to take on the role of 'teachers' for themselves and others. This challenges traditional power structures, making students more responsible for their own engagement in the learning process. It also creates a critical pedagogy based on game design and game creation, rather than the relatively passive learning of content through a gamespace predesigned by an authoritative instructor. Imagination is privileged as learners escape outmoded models of education, instead exploring the possibilities arising from the activation of imagination through digital gaming.

As Livingstone (2009) suggests, as long as the dominant ideology underpinning education continues to privilege hierarchical teacher–student relations, formalised testing and limited, instrumental specifications of learning outcomes, the educational potential of technology is likely to be circumscribed. Thus, the dissembling and challenging of rigid boundaries around school-based learning offer an opportunity for lasting educational gains. Not only will confronting traditional educational boundaries offer learning opportunities for students, but it could also lead to broader changes in teaching. After all, it can be argued that the apparent failure of classroom technology use to revolutionise learning reflects, at least partly, the limited expectations of educators.

Rieber (1996) observes that learning through designing is an excellent way for individuals to explore knowledge in rich and meaningful ways, promoting the concept of 'questions over answers' (see Khaled in this volume). Indeed, he suggests that 'the design process provides students with a relevant context for adapting content for a useful purpose… Rather than designing computer-based materials (and other forms of instructional technology) for children, perhaps a better strategy is to give them access to the most powerful design tools for them to use in their own design projects' (Rieber 1996: 52). This reflects the old adage that the most effective way to learn something is through teaching it to others. Commercial games such as the *LBP* franchise provide an opportunity for students to engage in such design-based learning. Even if the 'intended audience' fails to learn anything from the project, those who designed the game will know more about the content from the process of designing and building. Rieber (2001) goes as far as to suggest that school curricula should be based around design activities. He identifies that the key characteristics of such an approach would include work done largely in design teams, which consist of a diversity of people and expertise, individuals developing their own situated proficiency through an apprenticeship model and students learning to critically analyse both their own work as well as that of others (Rieber 2001: 7). This is precisely the model of High Tech High, a Charter School in San Diego which is completely built on the principle of Project-Based Learning.

As has been argued, key to GBL is the process of feedback. An educational approach privileging learning through game design would create new problems in terms of informal feedback and assessment. Yet this is in itself is not a reason to neglect this approach, particularly when the centrality of the notion of play is considered, with the opportunity to move beyond play as progress, power, fantasy or self, instead dissembling boundaries between teachers and learners, designers and players, creation and recreation. The nature of feedback and assessment would need to change with design-orientated learning, but this is necessary if an effective digital learning environment is to be realised.

7 Theorising *Little Big Learning*

In addition to the practical and pedagogical thrust of this paper, the term Little Big Learning itself also has further mileage in offering a potential critical theoretical framework. Given the gaming associations, the 'Little Big' moniker can be seen as signifying play and the creation of user-generated content. It certainly links with notions of play as progress, fantasy and self, but it could also be taken further. The 'Little' element indicates what has been identified as the traditional approach of GBL, as a method for delivering curriculum content through computer games. Yet, such views represent a narrow definition of learning, one that ignores the potential educational future anticipated by Rieber (2001) in which student design teams develop their own proficiencies through an apprenticeship model.

On the other hand, the 'Big' element of this proposed model might refer not only to a broadening of the notion of learning, but also to a dissembling of the tight boundaries often imposed by school curricula, and a resultant challenge to educational institutions who rely on polarised power relations between teacher and student. Although the 'Learning' element will have some distinct advantages over more traditional approaches, including playful, peer learning elements, it will also generate distinct challenges. This type of learning could also suggest a movement away from task-orientated educational processes, which inhibit the acquisition of broader knowledge and deeper understanding. Robinson (2009) notes that students who embrace a task-orientated approach in Internet use, avoiding activities that are not directly associated with school work, develop a 'taste for the necessary'. This framing of educational activity 'in terms of waste avoidance... ultimately does them a disservice, making it harder for them to develop more sophisticated information-seeking skills' (Robinson 2009: 505). In contrast, informal learning through 'playing seriously', a kind of enriching recreation, allows time to develop advanced abilities. Thus, adopting a less narrow approach to learning and providing space for students to 'play seriously' could benefit students. Insofar as 'Little Big Learning' would inevitably represent a loosening-up of task-orientated schooling, it could lead to substantial educational gains.

'Little Big Learning' might sound enticing as a fledgling conceptual framework, but what of the practical requirements of an education system built on rigid divisions of knowledge, traditional teaching approaches and the centrality of summative assessment? As discussed above, the need for formalised assessment and feedback may create difficulties for this playful approach to learning. Perhaps looking to other models of learning, such as project-based learning, might provide useful insights, but this belies the current reality of formal education in the majority of the economically developed world. Perhaps herein lies the greatest challenge to this model: a little idea that would require a big change not only in schooling practices, but also in the ideology that underpins them. This is truly playing on 'the edge' (see Cermak-Sassenrath in this volume, 'Playing on the Edge'), in ways that may seriously destabilise (or even undermine) our most treasured forms of educational institutionalisation.

Nevertheless, students are already designing user-generated content, playfully learning using technology in a savvy, even blasé manner. To ignore this reality would be a lost opportunity to capitalise on such learning. Failure to engage with how students learn in contemporary society might result in frustrating educational experiences, but given the constant possibility of resistance, it could also give rise to the hijacking of traditional pedagogies and curricula as students reconfigure the algebra of learning. Such challenging of boundaries is inherent in many forms of play, offering opportunities not only to develop new ways of knowing, but also seeking to 'democratise' educational practices through further empowering learners. This could reconstruct playful learning as central to educational progress, whilst forming part of a project of the self, (re)colonising learning.

8 Conclusion

It is becoming more widely accepted that game-based learning has a lot to offer formal educational environments. However, if GBL is only conceived of as the provision of tools (games) that are well designed by expert educators for passive learners to learn from, this is still a rather traditional, transmission-based model of education. To truly harness the potential of GBL, educators need to give the power of playful creation directly to students and allow them to build environments and artefacts with appropriate facilitation. This directly subverts some of the most entrenched social and power relations in educational institutions, including pedagogical relationships between and among teachers and students, as well as notions of feedback, assessment, design and the very process and product of learning itself.

Drawing upon anthropological notions of play is one way to critically reconsider the very idea of how learning is most effectively achieved. In this context, commercial products such as *LBP*, with their sophisticated user creation tools, could offer key learning spaces for the twenty-first century. Central to this process could be the notion of Little Big Learning, with all the little big challenges that come with it.

References

Arnseth, H. C. (2006). Learning to play or playing to learn—a critical account of the models of communication informing educational research on computer gameplay. *Game Studies: The International Journal of Computer Game Research, 6*(1). Retrieved May 30, 2010, from http://gamestudies.org/0601/articles/arnseth.

Behrens, J. T., Frezzo, D. C., Mislevy, R. J., Kroopnick, M., & Wise, D. (2007). Structural, functional, and semiotic symmetries in simulation-based games and assessments. In E. L. Baker, J. Dickieson, W. Wulfeck, & H. F. O'Neil (Eds.), *Assessment of problem solving using simulations* (pp. 59–80). New York: Erlbaum.

Biggs, J. (1999). *Teaching for quality learning at university: What the student does.* Buckingham, Philadelphia: Open University Press.

Blanchard, K., & Cheska, A. (1985). *The anthropology of sport: An introduction.* Massachusetts: Bergin & Garvey Publisher Inc.

Charsky, D. (2010). From edutainment to serious games: A change in the use of game characteristics. *Games and Culture, 5,* 177–198.

ConnectED (2013). LittleBIGPlanet2: Improving your STEM skills! ConnectED. Retrieved July 30, 2013, from http://www.connectededucation.com/products-services/sony-playstation/littlebigplanet-2-in-education/.

Crisp, G. (2007). *The e-Assessment handbook.* New York: Continuum International Publishing Group.

Csikszentmihalyi, M., & Csikszentmihalyi, I. (1990). *Flow: The psychology of optimal experience.* New York: Harper and Row.

Gee, J. P. (2007). *What video games have to teach us about learning and literacy* (2nd ed.). New York: Palgrave Macmillan.

Grant, S. (2011). Game on! Learning S.T.E.A.M. through LittleBigPlanet: Welcome to Discovery, Inc. *HASTAC: Humanities, Arts, Science and Technology Collaboratory.* Retrieved July 30, 2013, from http://www.hastac.org/blogs/slgrant/game-learning-steam-through-littlebigplanet-welcome-discovery-inc.

Ingold, T. (1994). Introduction to culture. In T. Ingold (Ed.), *Companion encyclopaedia of anthropology.* London: Routledge.

Johnson, L., Adams Becker, S., Cummins, M., Estrada, V., Freeman, A., & Ludgate, H. (2013). *NMC horizon report: 2013 higher education.* Austin, Texas: The New Media Consortium.

Jones, K. (1998). Simulations as examinations. *Simulation and Gaming, 29,* 331–341.

Krathwohl, D. (2002). A revision of Bloom's taxonomy: An overview. *Theory Into Practice, 41*(4), 212–218.

Lave, J., & Wenger, E. (1991). *Situated learning: legitimate peripheral participation*. Cambridge: Cambridge University Press.

Livingstone, S. (2009). *Children and the Internet*. Cambridge: Polity Press.

Pellegrini, A. D. (Ed.). (1995). *The future of play theory: A multidisciplinary inquiry into the contributions of Brian Sutton-Smith*. Albany, NY: State University of New York Press.

Rieber, L. P. (1996). Seriously considering play: Designing interactive learning environments based on the blending of microworlds, simulations, and games. *Educational Technology Research and Development, 44*(2), 43–58.

Rieber, L. P. (2001). Designing learning environments that excite serious play. In *Proceedings of the Annual Conference of the Australian Society for Computers in Learning in Tertiary Education (ASCILITE)*. Melbourne, Australia. Retrieved December 9–12, 2001, from http://www.nowhereroad.com/seriousplay/rieber-ascilite-seriousplay.pdf.

Robinson, L. (2009). A taste for the necessary: A bourdieuian approach to digital inequality. *Information, Communication and Society, 12*(4), 488–507.

Schechner, R. (1994). Ritual and performance. In T. Ingold (Ed.), *Companion encyclopedia of anthropology*. London: Routledge.

Shaffer, D. W., Hatfield, D., Svarovsky, G. N., Nash, P., Nulty, A., Bagley, E., Frank, K., Rupp, A. A., & Mislevy, R. (2009). Epistemic network analysis: A prototype for 21st-century assessment of learning. *International Journal of Learning and Media, 1*(2).

Shute, V. J. (2008). Focus on formative feedback. *Review of Educational Research, 78*(1), 153–189.

Shute, V. J., Rieber, L. P., & Van Eck, R. (2011). Games and learning. In R. A. Reiser & J. V. Dempsey (Eds.), *Trends and issues in instructional design and technology* (3rd ed., pp. 321–332). Boston: Pearson Education, Inc.

Shute, V. J., Ventura, M., Bauer, M., & Zapata-Rivera, D. (n.d.). Melding the power of serious games and embedded assessment to monitor and foster learning: Flow and grow. Retrieved May 30, 2010, from http://21st-century-assessment.wikispaces.com/file/view/GAMES_Shute_FINAL.pdf.

Squire, K. (2002). Cultural framing of computer/video games. *Game Studies: The International Journal of Computer Game Research, 2*(1). Retrieved June 10, 2010, from http://www.gamestudies.org/0102/squire/.

Squire, K. (n.d.). *Video Games in Education*. Retrieved July 12, 2010, from www.educationarcade.org/gtt/pubs/IJIS.doc.

Squire, K., & Jenkins, H. (2003). Harnessing the power of games in education. *Insight, 3*, 5–30.

Sutton-Smith, B. (1995). Conclusion: The persuasive Rhetorics of play. In A. D. Pellegrini (Ed.), *The future of play theory: A multidisciplinary inquiry into the contributions of Brian Sutton-Smith* (pp. 275–295). Albany, NY: State University of New York Press.

Sutton-Smith, B. (1997). *The ambiguity of play*. Cambridge, Mass: Harvard University Press.

Turner, V. (1983). Body, brain and culture. *Zygon, 18*(3), 221–246.

Whitton, N. (2010). *Learning with digital games: A practical guide to engaging students in higher education*. London: Routledge.

Woods, S. (2004). Loading the dice: The challenge of serious videogames. *Game Studies: The International Journal of Computer Game Research, 4*(1). Retrieved May 30, 2010, from http://www.gamestudies.org/0401/woods/.

Author Biographies

Chad Habel is an academic developer at Teaching Innovation Unit at the University of South Australia. His research interests include higher education, access and equity, curriculum design, educational technology, game studies and game-based learning. He has recently emerged from his dark age and is now a trained facilitator in the LEGO Serious Play method.

Andrew Hope is an Associate Professor in the School of Social Sciences at the University of Adelaide where he teaches sociology. His research interests include critical explorations of educational technology, surveillance in late modernity, and cultural criminology.

Subversive Gamification

Mathias Fuchs

Abstract Since the beginning of this decade, *Gamification* has become a buzzword for marketing, advertising and behavioural management, but also an accurate description of a fundamental shift in modern society: "Gamification is the permeation of society with methods, metaphors and attributes of games" (Fuchs 2012). Graphic game design elements, rule structures and ludic interfaces are exceedingly used by corporations to create and manage brand loyalty and to increase profits. This chapter aims at stirring up common sense notions of gamification as a marketing tool and will discuss alternative artistic concepts, activist tactics and subcultural strategies aiming at a subversive ludification of society.

1 Introduction

Today we encounter a vast offer of gamified applications that promise to strengthen customer loyalty, to increase profit or to create other benefits for users and society. But there are currently only few attempts to apply gamification mechanics in a critical and subversive manner. It seems however essential for an understanding of the societal effects of gamification, to take a closer look at the rather controversial and less general aspects of playfulness. Core objects of the analysis are artistic interventions, playful hacking and ludic disobedience. Gamification has been ridiculed as a mere buzzword, but it is also a symptom of an underlying, fundamental transformation of our society. The trendy term is not embraced warmly by everybody. Ian Bogost's remark that "Gamification is bullshit" (Bogost 2011), Dragona's concept of "counter-gamification" (Dragona 2013), Escribano's *Mene Tekel* of a "ludictatorship" (Escribano 2013) oppose the emphatic use of the notion and will be laid out in this article. Concrete apps, games and interventions will be looked at with a critical eye. A comparative analysis of the "destroy all surveillance

M. Fuchs (✉)
Leuphana University, Gamification Lab, Lüneburg, Germany
e-mail: mathias.fuchs@creativegames.org.uk

© Springer Nature Singapore Pte Ltd. 2018
D. Cermak-Sassenrath (ed.), *Playful Disruption of Digital Media*,
Gaming Media and Social Effects, https://doi.org/10.1007/978-981-10-1891-6_12

cams" game, that Berlin anarchists promoted as playful political action, and artistic interventions leads to a distinction of different modes of subversiveness: The chapter suggests that there is a difference between the subversive rhetorics of gamification, and political subversion in a playful manner.

2 Subversive Gamification

I have suggested elsewhere that gamification can be understood as a new form of ideology, a form that might even be the dominant form of ideology in the twenty-first century (Fuchs 2014). When the evangelists of gamification tell us that work should be enjoyed as being playful, that our personalities are experimental avatars, that the whole economy is nothing but a game and that each and every activity from cradle to grave can be turned into play, we encounter false consciousness that is socially necessary. Today gamification is used to tell people that if reality is not satisfactory then at least play might be so. Jane McGonigal phrased this aptly in her popular proposal that "Reality is broken" (McGonigal 2011). Replacing reality-based praxis with storytelling, gaming, self-motivation or "self-expansion escapism"[1] is what Marx and Engels would have labelled "ideology". McGonigal's "When we're playing games, we're not suffering." McGonigal (2013) is the cynical statement of a writer/designer/self-promoter who is definitely not suffering economically and has little reason and even less time to play games. But ideological and necessary false statements on the relationship of work and play have not been premiered in the current decade. They were invented and intentionally introduced as ideology before computer games came into existence.

In 1934, Pamela Lyndon Travers had her famous novel's main character Mary Poppins say:

In ev'ry job that must be done

There is an element of fun

You find the fun, and snap!

The job's a game! (Travers 1934)

Such a statement, uttered in 1934, was far from true or poetic but rather a cold-blooded proposition to smile, when the drudgery of labour was unbearable for a large part of the population. The lines were conceived in times of economic crisis and pre-empted the *gamification* evangelicalism of our days. It was only 5 years after the *Black Thursday* of 1929 when Mary Poppins is suggested to consider work fun. Almost a century later, the notion of gamification was introduced widely (Zichermann and Cunningham 2011; McGonigal 2011; Deterding et al. 2011a; Schell 2011) to suggest that marketing, warfare, health and labour might be some

---

[1]http://www.polygon.com/2013/3/28/4159254/jane-mcgonigal-video-game-escapism.

kind of free play or leisure activity. This was just a few years after the so-called credit crunch deprived many of work. This chapter analyses the controversial dialectics of play and labour and the ubiquitous notion of *gamification* as ideology. The question is raised here, whether the affirmative process aiming at total gami-fication of society has a counterpoise of subversive gamification. Subversive gamification would be an apparition (Adorno 1984, p. 104) or a glimpse of hope in a situation that has been described as "ludictatura" or "ludictatorship" (Escribano 2013). Subversive gamification can be seen as a strategic move, an aesthetic operation or a rhetoric figure to oppose the totality of playfulness.

There are two coinciding reasons that convince us to conceive gamification as ideology[2]:

1. Gamification is false consciousness: Gamification promises to offer a method that could make work compatible with self-realisation and fun. The proposition that game design elements can change the nature of labour and successfully cope with exploitation, "alienation" (Zichermann and Linder 2013) or "suffer-ing" (McGonigal 2013) is proven on the basis of subjective assessment or mere speculation, and not based on empirical economic analysis.
2. Gamification is socially necessary: Concluding from market analysis and market predictions data that Ernest and Young,[3] Saatchi and Saatchi and Gartner[4] offer, business needs to implement gamification in most of the sectors that drive our economy. The reason for that is that according to capitalist logic, economic sectors have to grow. Gamification methods are seen as a means to avoid stagnation of productivity by keeping the customers (and the workforce) satis-fied. It will therefore be mandatory for consumers and prosumers, i.e. consumers that voluntarily or involuntarily contribute to the production of commodities, to embrace gamification as well. Gamification is not a choice, it is socially necessary.

Ideology works best when it distorts reality in such a way that we do not notice the distortion, because everything seems to be alright. While in fact a mistaken identity and a unification of play and labour serve the needs of the economic system, the ideas of ideology make it appear natural. Althusser observes too dif-ferent kinds of what he calls state apparatuses: the repressive state apparatus, i.e. is the military and police, and the ideological state apparatuses that have formerly been religion, education, family, sports, culture. The repressive state apparatus functions "by violence", whereas the ideological state apparatuses function "by ideology" (Althusser 1971). Ideology is the soft and yet the most effective way of executing subordination. It makes the subordinate classes accept a state of

---

[2]As Joseph McCarney demonstrates in his text on "Ideology and False Consciousness" (2005) Marx never talked of ideology as "false consciousness". Althusser and Sohn-Rethel however drew a line between ideology and false consciousness much later.

[3]Ernest and Young (2011). 5 things you need to know about gamification. http://performance.ey. com/wp-content/uploads/downloads/2011/11/EY_performance_Review_pg28_Ideas.pdf.

[4]Brian Burke for Gartner (2012) Gamification: Engagement Strategies for Business and IT.

alienation against which they would otherwise revolt. This state of alienation is stabilised by ideology that looks completely natural on the surface. In our days religion, education and family have lost their ideological potential; they look a bit worn out and are not accepted as "natural" guiding principles for behaviour. It is therefore now the time to replace these ideological frameworks with a new one, that everybody likes: Gamification. In the closing chapter of Alfred Sohn-Rethel's "Intellectual and Manual Labor" (Sohn-Rethel 1978), he invokes the concept of "necessary false consciousness". This is a type of consciousness that is not just faulty consciousness. Necessary false consciousness is rather an ensemble of ideas, legitimisation mechanisms and moral codes that is logically flawless. For the very reason of the inherent logic of ideology, ideology cannot be proven to be logically inconsistent. It can only be subverted.

3 Subversive Gamified Activism

In 2013, various media reported about a game to destroy CCTV cameras that activists in Berlin have developed to gamify the process of destroying surveillance devices. In this game points are given to the players for smashing cameras, with bonus scores for the most innovative modes of destruction (Fig. 1).

"As a youth in a ski mask marches down a Berlin U-Bahn train, dressed head-to-toe in black, commuters may feel their only protection is the ceiling-mounted CCTV camera nearby. But he is not interested in stealing wallets or iPhones—he is after the camera itself. This is Camover, a new game being played across Berlin, which sees participants trashing cameras in protest against the rise in close-circuit television across Germany.

The game is real-life Grand Theft Auto for those tired of being watched by the authorities in Berlin; points are awarded for the number of cameras destroyed, and bonus scores are given for particularly imaginative modes of destruction. Axes, ropes and pitchforks are all encouraged".[5,6]

Camover is a form of resistance that highlights the "importance of the power to act against the power" as Daphne Dragona puts it, (2014, p. 238) following a proposal of Gilles Deleuze. Dragona sees counter-gamification related to the concept of Gilles Deleuze's "counter-actualisation"[7] that highlighted the possibility of one becoming the actor of his own events. She also relates counter-gamification to

---

[5]www.theguardian.com/theguardian/shortcuts/2013/jan/25/game-destroy-cctv-cameras-berlin.

[6]The rules of Camover are such: Mobilise a crew and think of a name that starts with "command", "brigade" or "cell", followed by the name of a historical figure. Then destroy as many CCTV cameras as you can. Finally, video your trail of destruction and post it on the game's website.

The competition ends on 19 February, to coincide with the start of the European Police Congress. The prize is a front-row position in a rally against police violence in Berlin.

[7]Deleuze (1999). *The logic of sense*. Transl. by Mark Lester with Charles Stivale. Ed. by Constantin Boundas. New York: Columbia University Press, pp. 150–161.

Fig. 1 Screenshot from a movie about activists destroying CCTV cameras in Berlins's subway

Hardt and Negri's "counter-empire"[8] that emphasises the potentiality of multitude for resistance. Dragona suggests that counter-gamification could use any of the tactical approaches of obfuscation, over-identification, hypertrophy, exodus or devaluation to name just a few (Dragona 2014, pp. 237–245). Each of these tactical moves can be political. The case of Camover lends itself as an example for devaluation and obfuscation. The former is accomplished by trashing the surveillance cameras, and the latter one is accomplished as the players devise fake names for their teams. A name like "Brigade Rosa Luxemburg" is in this case not adopted for the reason of identification with a historic figure, but rather in order to create confusion and camouflage. Counter-gamification differs from traditional political action as names and meaning of terms are often introduced playfully. The political action staged by the Camover team—the name of the game is a pun in itself—intends to stir up a situation with humour and playfulness. The guideline for counter-gamification of this kind is therefore not the seriousness of well-organised class struggle, but a political and hedonistic attitude that follows the motto of "every event or demonstration should be planned so as to be fun for the participants".[9]

---

[8]Negri and Hardt (2000). *Empire*. Cambridge Masachussets: Harvard University Press, pp. 205–218.

[9]Wolfgang Lefèvre quoted from Walther (2008). *Ein direkter Weg von der Spassguerilla zum Terrorismus? Aktions- und Gewaltformen in der Protestbewegung. 68: Jahre der Rebellion.* Bundeszentrale für politische Bildung. www.bpb.de/geschichte/deutsche-geschichte/68er-bewegung/51795/spassguerilla-terrorismus.

4 Aesthetic Subversion of Gamification

Gamification can be a tool, a *sujet* or a strategy to criticise the ideological character of gamification. In his project "Start-A-Revolution",[10] the artist Friedrich von Borries in collaboration with designer Mikael Mikael and the artist Slavia appropriates the jargon, visual appearance and the game mechanics of gamified apps and social networks to call for participation in starting a revolution. Von Borries and friends dress an absurd proposition—that the revolution can be gamified—in a costume of social media chique and website mimesis. We seem to have become so much accustomed to Web 2.0 mechanics that the artist can present a "Resistance Ticker", badges for revolutionary activities, actual challenges and the notorious "Thumbs Up" icons on a Web page in close vicinity with Facebook and Twitter links to make the page look completely plausible.

The statement that Friedrich von Borries' group makes here is inherently logical. If we believe that everything in the world can be gamified, then the revolution must as well be an object of gamification. The artists' project is much more than a mockery of hyped icons, services and notions. Von Borries tries to subvert an ideology by showing that this ideology is false and at the same time necessary in the sense Sohn-Rethel conceived ideology to be "necessarily false" (Fig. 2).

The name RLF[11] for the project turns Adorno's famous memento from *Minima Moralia* that "there is no right life in falsehood" ("Es gibt kein richtiges Leben im falschen" (Adorno 1980, p. 43)) off its head, and places it upon its feet and into the artists' shoes. The RLF Web page offers shoes as well, as any Web page nowadays seems to sell T-shirts and gadgets. The RLF shoes are exclusively redesigned limited edition adidas sneakers of the type "Adidas Torsion Allegra X". The shoes contain pure gold decorative elements and a revolutionary message embedded invisibly in the sole that will create footmarks after the soles wore off. The artists promoting the shoes advertise it with the words "this shoe cannot be bought, it has to be fought for" ("Kann man nicht kaufen. Nur erkämpfen."). All of the game elements like challenges, points and badges contribute to a consistent experience of gamification, yet the content is compromised by the ideological framing. For the very same reason that alienation of labour and alienation of workers cannot be solved via religion, revolution cannot be accomplished via gamification. By showing us how gamification would shape such processes as revolutionary praxis the artist makes us aware of the difference of play and politics in a playful manner. In other words, gamification is a rescue mechanism for a reality that is said to be broken. The distorted view on this reality is subversively attacked.

Another example of play on an aesthetic level in the age of the ubiquity of computer games is the artwork of Michael Johansson. Johansson uses piled up cars,

---

[10]www.rlf-propaganda.com/.

[11]Adorno would hardly have advocated to abbreviate his one-liner of the right life in falsehood into the a three-letter word of RLF and thereby put it in a league with BMW, NATO or VW.

Fig. 2 Screenshot from the Start-A-Revolution Web page by Friedrich von Borries

furniture and refrigerators to create bizarre photographs of a hypothetical gamified, extraordinary life.

Johansson refers to the rules he observes in computer games, to gaming stereotypes, to entertainment classics such as Tetris and to the gameplay imperatives that do not stay within the "sphere of games" (Huizinga 1949, p. 17), but escape the magic circle and change our world. When Johansson shows to us stacked cars, piles of books and containers in colourful geometric arrangements, we become aware of the fact that we constantly encounter the use of "Game Design Elements in Non-Gaming Contexts" (Deterding et al. 2011b). The artist picks up the viewpoint and definition Sebastian Deterding et al. introduced, but his artistic statement is not affirmative of gamification. As an artist he has not to be bothered about design processes getting more effective, user-friendly or easily accepted. For Johansson, gamification seems to be an aesthetic issue and his surreal large-scale assemblages comment on the myth of computer games in a subversive and gamifyed manner.

5   Subversive Rhetorics of Gamification

Gamification is a term that has been appropriated by the wielders of power, if they have not coined the term themselves.[12] The rhetorics of gamification consist of a *promesse de bonheur* ("health", end of "suffering", "self-motivation", "self-expansion" (McGonigal 2011)), of totalitarian threats ("Games are the new Normal" (Al Gore 2011)[13]) and of monetary incentives ("Gamification is projected to be a \$2.8 billion market by 2016" (M2 Research 2013)[14] or "\$5.8 billion for 2018" (Markets and Markets 2013)). The rhetorics of gamification resemble forms of discourse of other ideological systems. When Protestant preachers of the eighteenth century used *pathos*, *logos* and *ethos* to pull all strings of persuasiveness, they usually mixed promises (paradise) with threats (hell) and with monetary incentives (economic growth) (Weber 2005).

The question we want to raise here is whether this affirmative rhetoric can be counter-balanced by a critical rhetoric about gamification and within gamification contexts. One of the brilliant attempts to do so is Flavio Escribano's neologism of "ludictatorship" (Escribano 2013). To combine the antagonistic notions of dictatorship and ludus means to subvert the ideological idea that games are a "free activity" (Huizinga 1949, p. 17) and to direct a problem discovered on the level of *logos* to an effect on the level of *pathos*. Dictatorship is a term that implies fear and evokes discomfort. Other then with traumatic events from the past like the 1936 Olympics in Berlin, ludictatorship suggests that it is not the bad guys using games amongst other things for their purposes right now, but that the total regime of playfulness turns into a totalitarian concept. Al Gore's statement about "the new Normal" (Al Gore 2011) is strongly suggestive of such a situation. If games are the normal, how do non-gaming activities qualify then? Certainly non-normal, probably pathologic, maybe dispensable.

Ian Bogost seems to be of the opinion that gamification is dispensable, when he talks about gamification as "Bullshit" (Bogost 2011). He designed and made freely

---

[12]It is interesting to see that the term of a "ludification" that was introduced by Joost Raessens as early as Raessens (2006) has never managed to get the popularity that the 2010s term of *gamification* immediately got. This might have to do with the connotation of *ludification* with the cultural sector and with the smell of big bucks that *gamification* had from the start when the likes of Zichermann, Saatchi and Saatchi or Ernest and Young dropped the bomb of a new alchemistic preciosity with the name of gamification.

[13]www.psfk.com/2011/12/al-gore-games-are-the-new-normal.html.

[14]m2research.com/Gamification.htm.

Fig. 3 Screenshot from Ian Bogost's Cow Clicker game

available an anti-game called "Cow Clicker",[15] that he describes as a Facebook game, but that could well be called a gamified app (Fig. 3).

Bogost tells us: "Cow Clicker is a Facebook game about Facebook games. It's partly a satire, and partly a playable theory of today's social games, and partly an earnest example of that genre.

You get a cow. You can click on it. In six hours, you can click it again. Clicking earns you clicks. You can buy custom "premium" cows through micropayments

---

[15]www.bogost.com/blog/cow_clicker_1.shtml.

(the Cow Clicker currency is called "money"), and you can buy your way out of the time delay by spending it. You can publish feed stories about clicking your cow, and you can click friends' cow clicks in their feed stories. Cow Clicker is Facebook games distilled to their essence." (Bogost 2011). Bogost's game is gamification distilled to its essence. The main thing is to gain points, to spend money and to be kept in the process of continuing the game. The game that is supposed to be satirical does actually not differ very much from gamified apps that are dead serious. In this regard, it resembles many acts of protest that share the channels and the idiom of the oppression. Bogost's rhetorical figure is parody. He does not accuse or criticise on the level of *ethos* but subverts gamification by showing that even such an extreme app like Cow Clicker can be played and will be played. Bogost thinks that games are corrupted through gamification and describes it as "marketing bullshit, invented by consultants as a means to capture the wild, coveted beast that is videogames and to domesticate it for use in the grey, hopeless wasteland of big business, where bullshit already reigns anyway".[16] Bogost does however not think that the makers of gamified apps are stupid. "Bullshitters are many things, but they are not stupid. The rhetorical power of the word 'gamification' is enormous…".[17]

On the level of rhetoric manoeuvres there is also an ongoing struggle about terminology taking place. Raessen's *ludification* from 2006 (Raessens 2006), the anglo-american *gamification*, as well as a whole set of European flavours and regional claims put into the grounds of the discursive fields that ask for differentiating between παιγνιδοποίηση, *Ludicizzazione*, *Ludificação*, *Gamificación*, *Ludización* or "*Ludifizierung*" contribute to the process of subverting the idea that there is one type of gamification—and one type only. This is helpful in deconstructing a concept that could easily appear to be exclusive. Subversive rhetorics could not only challenge whether gamification is good or bad, it could even differentiate between economically motivated *gamification apps*, *gamification with a purpose*, *critical ludification*, pedagogically useful "*Ludifizierung*", ideological gamification and *subversive gamification*.

If subversive gamification is at all possible, we might come closer to developing methods and tools for this very subversive gamification, by acknowledging that there is a difference in between the subversive rhetorics of gamification, subversive aesthetics and political, subversive gamified action.

References

Adorno, T. W. (1995 = 1970, engl. 1984). *Ästhetische Theorie*. Frankfurt/M.
Adorno, T. W. (1980 = 1951, engl. 1974). *Minima Moralia: Reflections from damaged life*, (E. Jephcott, Trans., *Minima Moralia: Reflexionen aus dem beschädigten Leben*). Frankfurt am Main: Suhrkamp.

---

[16]www.bogost.com/blog/gamification_is_bullshit.shtml.
[17]Ibid.

Althusser, L. (1971). Lenin and philosophy and other essays (B. Brewster, Trans.). In *Ideology and ideological state apparatuses*. Monthly Review Press.

Bogost, I. (2011). *Gamification Is Bullshit! My Position Statement at the Wharton Gamification Symposium*. In: Ian Bogost Blog. Retrieved August 8, 2011, from http://www.bogost.com/blog/gamification_is_bullshit.shtml.

Deterding, S., Khaled, R., Nacke, L. E., & Dixon, D. (2011a). *Gamification: Toward a definition*. Vancouver.

Deterding, S., Dixon, D., Khaled, R., & Nacke, L. (2011b). From game design elements to gamefulness: Defining "gamification". In *Midtrack Conference*.

Dragona, D. (2014). Counter-gamification. Emerging forms of resistance in social networking platforms. In M. Fuchs, S. Fizek, N. Schrape & P. Ruffino (Eds.), *Rethinking gamification*. Lüneburg: Hybrid Publishing.

Escribano, F. (2013). Gamification versus ludictatorship. From sex games to Russian roulette. In *Gamfication Conference at Leuphana University*, 15 May 2013 (unpublished).

Huizinga, J. (1949, Dutch original 1938). *Homo ludens. Homo Ludens: A Study of the Play-Element in Culture*. Routledge.

McGonigal, J. (2011). *Reality is broken: Why games make us better and how they can change the world*. New York: The Penguin Press.

Raessens, J. (2006). Playful identities, or the ludification of culture. In *Games and culture* (Vol. 1, no. 1, pp. 52–57). Sage Publications.

Sohn-Rethel, A. (1978). *Intellectual and manual labour: Critique of epistemology*. Macmillan.

Travers, P. L. (1934). *Mary Poppins*. London: G. Howe.

Weber, M. (2002, German original 1905). *The protestant ethic and the spirit of capitalism*. US: Penguin Group.

Zichermann, G., & Cunningham, C. (2011). *Gamification by design: Implementing game mechanics in web and mobile apps*. O'Reilly.

Zichermann, G., & Linder, J. (2013). *The gamification revolution: How leaders leverage game mechanics to crush the competition*. McGraw-Hill Professional.

Author Biography

Mathias Fuchs (* October 20, 1956 in Erlangen/Germany), studied computer science in Erlangen and Vienna (Vienna University of Technology), and composition in Vienna (Universität für Musik und darstellende Kunst Wien) and in Stockholm (EMS, Fylkingen). He has pioneered in the field of artistic use of games and is a leading theoretician on Game Art and Games Studies. He is an artist, musician, media critic and works at Leuphana University/Lüneburg. He is also director of the Leuphana Gamification Lab.

Constant beyond Gamification: Deep Play in Political Activism

Margarete Jahrmann

Abstract Playful practices in historic and contemporary forms of political "activism" had a traceable impact in the formation of political consciousness and identity in everyday life. The following analysis reflects the social implications of such political ideas about play as principle and follows trajectories of political agency through a close look at the author's work as game artist in her project *Ludic Society* and the play with identity (mimicry), performance, and creative practice in social and arts avant-garde experiments. It compares the ethnographic concept of *Deep Play* (Geertz, The interpretation of cultures, theory of culture 1973) with current concepts of activist role play, social intervention, and public protest against certain conditions of work, society, and urbanity. The chapter finds its creative and intellectual leitmotif in "ludic" activist arts connected to contemporary forms of game arts and political role play. Its claim for the efficacy of such ludic practices is informed by the theoretical concept of *Deep Play*.

1 Notes on Deep Play and a Social Playfulness Discourse

Study *Deep Play* by the sociologist Geertz (1973) underlines this chapter's argument about playfulness as a political vehicle for activism and agency. Geertz's original observation was informed by interpretations of the effects of betting and winning as motivation mechanics of play on social status, and how a play result deeply affects the player's position in a social hierarchy. The concept of social hierarchies and *Deep Play* was applied as a particular methodological objectification from the broader context of anthropologies. It builds on the mutual information of investigated subject and research object and was introduced as a social analysis method with a case study of Balinese cockfighting as a particular play practice with

M. Jahrmann (✉)
Game Design, Zürcher Hochschule der Künste, Zürich, Switzerland
e-mail: margarete.jahrmann@zhdk.ch

M. Jahrmann
Digitale Kunst, Universität für angewandte Kunst Wien, Vienna, Austria

© Springer Nature Singapore Pte Ltd. 2018
D. Cermak-Sassenrath (ed.), *Playful Disruption of Digital Media*,
Gaming Media and Social Effects, https://doi.org/10.1007/978-981-10-1891-6_13

the highest social impact on a certain society. Geertz's analysis of an immersive betting game answers the question of how social hierarchies are embodied by play. Geertz observes and tries to understand players who bet on fighting cocks and through this game of chance gain or lose social reputation, money, and credibility. In this analysis, Geertz describes a particular situation of the embodiment of social relations through play as "What makes Balinese cockfighting deep is thus not money in itself, but what, the more of it that is involved the more so, money causes to happen: the migration of the Balinese status hierarchy into the body of the cockfight" (Geertz 1973: 441).

This view is supported by the identification of play as a frame for the formation of social status, which demonstrates how political and social consequences entirely outside play can be embedded in it. The observation of how a society integrates play principles into social life is indicated as status hierarchy (Geertz 1973: 444) in the study. Under this term, it details a play system as a vehicle to play out hierarchies, which gives insights in comparable play systems found in today's social networks. As Geertz demonstrates, play can be experienced as embodied social relation. According to this understanding of play, the subject's role in the framework of a socially informed exchange is elaborated in its symbolic dimension. The value of playful social interaction is introduced as an antithesis to a capitalist understanding of bet and win as a tool to generate individual surplus. Based on this insight, it is possible to identify a need to consider playfulness in relation to social hierarchies of everyday life. In contemporary worlds, this approach of revisiting playfulness as relational play on social status can subsequently be examined as inherent to the structure and requirements of technologically supported plays on status in social networks.

However, Geertz' intervention emphasizes play using the method of "thick description."[1] By definition, thick description is a very dense view of certain actions, close to an incorporation of the matter investigated. Under this angle the application of play in research results in the immersion of investigated research subjects and research hierarchies. It describes the dense involvement of the researcher in the phenomenon investigated, which results in the accuracy of the terminology developed in a discipline. Such involvement can be identified as my personal background as an artist involved with activism and urban games. Examples of such ludic art games can serve as a source for the argument of the deep entanglement of play and activism. In particular under the label *Ludic Society* (2006–2016) I developed pervasive, urban, and alternate reality games with an applied activist *Deep Play* aspect. The games (featured under the address www. ludic-society.net) aimed to touch certain elements of social status play in arts and games research communities. I introduced the name *Ludic Society* as a label for the

---

[1]Geertz compares the method of thick description of an interpretive anthropologist, who accepts a semiotic view of culture, with the method of the literary critic when analyzing a text: "Analysis, then, is sorting out the structures of signification—what Ryle called established codes—and determining their social ground or import" (http://academic.csuohio.edu/as227/spring2003/geertz. htm [Accessed: February 21, 2016]).

Image 1 VOIDBOOK presentation, *Ludic Society* founders, Cabaret Voltaire, Zürich 2016. *Foto* Ludic Society

development of a series of activist art games, which were accompanied by a corresponding series of theoretical publications. In its core the concept is informed by approaches to agency through play. In the peer-reviewed periodical *Ludic Society magazine* we published and featured between 2006 and 2016 a methodological approach of associative arts texts, including experiments such as automatic or 'pataphysical[2] writing instead of pure analytical texts'. Up to now the magazines contain articles of 49 international authors, who discuss ludic theories and ludic art. The magazine always comes in print, but pdf documents of the magazines are available online on the Ludic Society webpages. On the occasion of 10 years of *Ludic Society* we published and presented the VOID book on the topic of emptiness. This book was premiered in ludic soirées, including performance and play at various locations, such as the Cabaret Voltaire, the founding place of DADA in Zürich, or at conferences including the ISEA, Inter Society of Electronic Arts Hong Kong (2016) conference.[3] Documentation and manifestos on the VOID play performances, featuring the book, are available at www.ludic-society.net/voidbook (Image 1).

---

[2]'*Pataphysics*, a term coined by the French writer Alfred Jarry, is a philosophy dedicated to studying what lies beyond the realm of metaphysics. It is a parody of the theory and methods of modern science and is often expressed in nonsensical language.' Pataphysics is also defined by Jarry as the "science of imaginary solutions" (http://www.urbandictionary.com/define.php?term=pataphysics [Accessed: February 21, 2016]).

[3]Urban Screen, Public Art, The Ludic Society's Void Book (2016), https://isea2016.scm.cityu.edu.hk/openconf/modules/request.php?module=oc_program&action=program.php [Accessed: March 12, 2016].

Image 2 *Neural* magazine
cover, ludic art pieces/
Dangerous Games, 2008.
Foto Ludic Society

Crucial for political "activism" based on play in the Ludic Society project was a connection of game art-based works to a critical approach towards society, in particular in relation to conditions of technologically shaped everyday life and the role of the individual and her identity in such a technologically shaped society. The Italian art critic and media theorist Alessandro Ludovico (2008: 2) discusses the *Ludic Society* art project:

> The *Ludic Society magazine* involves different cultural sectors and perspectives of the analysis of the real. This magazine is a precious independent voice, striking a discordant note compared to the suddenly established academic "videogames studies." From the 'Pataphysics to the role and potential of the graffiti and tag in the videogame', until the "real game," or the game played in the public urban spaces, there's a vast and free editorial perspective. It is pointed in different directions but with a common horizon, and it is framed in a '90s zine layout, comic size, using striking black and white contrast. Here the "game rules" rise to the level of a vital paradigm, implicitly defining "ludology" as an ironic social life science.
>
> (*Technology and Cultures Magazine neural*, issue 30, http://www.neural.it/nnews/ludic_society_magazine_e.htm [Accessed: May 12, 2016]) (Image 2).

Issue 30 of the *neural* magazine was published with a cover featuring a *Ludic Society* art piece on electronic urban tagging. The image shows an absurd play interface, and connects it to the topic of "dangerous games." This publication provides evidence of the public media echo on such critical and absurd game art pieces. The attention and political discourse caused by the reviews of the magazine demonstrate the viability of the thesis of agency through play. Public attraction was achieved by the absurd coupling of game mechanics with play in urban streets, as space to achieve a broad public appearance.

The *Ludic Society* magazines featured political play and game art pieces. The issues have been exhibited as artifacts in museums since the first *Ludic Society* exhibition in *Neue Galerie Graz* 2006. The aesthetically appealing magazines with graphical art by Max Moswitzer became collectors' items. For example, in 2010, a

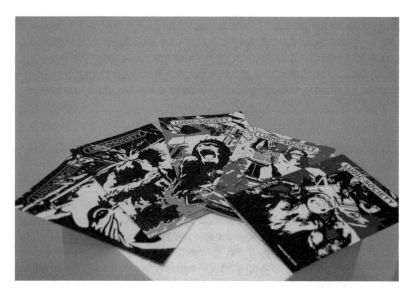

Image 3 *Ludic Society* magazine issues, 2006–2009. *Foto* imonym

reprint of *Ludic Society* issue 1 was published in the Swedish media theory journal *OIE*, discussing the playful writing project of the ludic magazines (see http://www.oei.nu/w/6.html) (Image 3).

Drawing from *Ludic Society* magazine excerpts from 2006 to 2008, the project can be described by the following: the Ludic Society, as an international association, exists to provoke an artistic research discipline best addressed as ludics. According to Friedrich Nietzsche's dictum of the gay science, it is prescribing laughter when talking about serious games studies. Inspired by Nietzschean tropes, lived ludic practice leads to a fully cheerful science. As an *ouvroir* (French: garage) for contingency and imaginative solutions, its methods are what the French would call *ludique*, which is to say playful, amusing, and, by extension, really rather puzzling. The foundation of an arts research association as arts project tests the hypothesis of anthropological play on social hierarchies as introduced above. The society was founded as unbloody status play. The affiliation was built systematically as an analogue social network structure of peers in a particular field of excellence around play. As political statement it was launched in opposition to electronic social networks with a magazine in print, real-life meetings, and chapters. The *Ludic Society* magazines were published as a periodical with articles from international authors, on tendencies of playful interventions in arts and theory. The black and white print in the style of a fanzine and the particularly chosen quality of recycled natural paper expressed a reference to fan cultures and their agency. As aesthetic artifact, the magazine was made strategically public along with joyful performances. Ludic research anticipated action research methods, because many members are active in game development, theory, and arts, and see no problem in defining their work as play investigation. The *Ludic Society's* mission was to provide a playful theoretical starting point for a methodology around the act of play as a state of transformation towards an activist consciousness. The main argument builds on the observed nascence of an experience-based emotional reflection, which is achieved by playing through game systems as a looking glass into everyday life.

2 An Exemplary Selection of Activist Ludic Society Game Art Works

The term *ludic* has an etymological root in *ludus*, which embraces *paidia* and *ludus* in the sense that *ludus*[4] translates, according to Lewis (1994) from Latin as both game and play. Ludics, as a method for arts and research, converges game and play to gain a clear understanding of the involved connections to agency and activist mechanics. Accordingly, ludic artwork exemplarily embraces a political dimension by offering practices of resistance to technological domination and regulation, which are socially established and practically enabled by a contemporary ubiquitous electronic realm. The arts practice applied in *Ludic Society* works demonstrates in the opposite way that interventionist arts in electronic networks can be shaped by theoretical interventions. This is relevant for the new emerging ludic societies[5] of the future. The *Ludic Society* art embraces play, technologies, and discourse critique. The art pieces published under the label of the *Ludic Society* are understood as experimental setups for testing the hypothesis of play as a vehicle of political agency when it comes to deal with technologically determined realities and its constituting conditions. As such, the *Ludic Society* pieces cover the range of absurd objects (*New Bachelor Machines – objets célibataires, 2007–2008*, http://ludic-society.net/play/objects.php), and the agency that is provoked by the absurdity of electronic objects and social networks. In this art piece, by inversion of introduced orders of technology and power structures, theoretical concepts of play are examined for their general viability for political agency in society (Image 4).

With the art performance *Evening of the Ludic Society* (Olli Leino, René Bauer, Max Moswitzer, Margarete Jahrmann 2007), as presented at the RO theatre Rotterdam at the Dutch Electronic Arts Festival 2007, RFID (radio frequency identification) technology was used for a subversive art play with mobile phones as

---

[4]*Ludus*: a play, game, diversion. In general, a play, game, diversion, pastime: "*ad pilam se aut ad talos, aut ad tessaras conferunt, aut etiam novum sibi aliquem excogitant in otio ludum,*" Cic. de Or. 3, 15, 58: "*datur concessu omnium huic aliqui ludus aetati,*" id. Cael. 12, 28: "*campestris,*" id. ib. 5, 11: "*nec lusisse pudet, sed non incidere ludum,*" Hor. Ep. 1, 14, 36.— B. In partic.1. Ludi, public games, plays, spectacles, shows, exhibitions, which were given in honor of the gods, etc. (Charlton T. Lewis, *A Latin Dictionary*, 1879).

[5]The term society is written in lower case in this instance, in order to address society as a social construct in general, as opposed to Ludic Society as arts project, which is introduced here as an art practice.

Image 4 objet célibataire, *artisan circuit board/ludic interface*, Jahrmann (2006). *Foto* Jahrmann

factual objects. In issue five of the *Ludic Society* magazine, which covers this artwork, the economist Edward Castronova (2005) describes the economic relation of the synthetic and the real as a process of play. He draws parallels from the political consequences of the economic crisis of the 1920s to the contemporary online play currencies of immaterial virtual worlds' economies (Image 5).

Image 5 Ludic Wheel performance, SESC, Avenida Paulista, Sao Paulo, 2007. *Foto* Ludic Society

Image 6 Ludic Wheel, sel-etched circuit board/'pataboard, Jahrmann (2007). *Foto* Jahrmann

The hybrid play concept of an aesthetically anthropomorphized technological object is tested in the artwork *Ludic Wheel*, 2007. It demonstrates that aesthetically shaped technological objects can become tools of economic and technological power (Image 6).

The *Ludic Society* monowheel is an absurd instrument in the 'pataphysical tradition; it is useless but at the same time a working technological tool, an artisan expressive circuit board, which serves as a live performance instrument in the Reality Game Engine. The self-etched circuit board, its intrinsic 'patapyhsical circuitry providing electric current rides on the wild side of copper are design metaphors, transgressing theory machines into an experience of printed circuit boards as "*objet de jeu, de vie et d'art*" (objects for play, life and art), whereas the monowheel itself remains as a self-sufficient game interface. Preserving the extravaganza of an absurd and useless interface, the ambiguity of the play object oscillates between retorsion of unspoken words and its chosen spiral slope metaphor, but never neglects its self-inherited uselessness, being a "reality engine device" to play and live the Dada-funk! (Image 7).

The artwork *Blitz Play Bergen* (Ludic Society 2008) integrates contemporary electronic toys as new kinds of ironic objects. The aim of this contemporary performance artwork with technologies is to fight surveillance dimensions in the electronic realm of wireless communication and interaction.

This arts practice provides the basis for a ludic methodology of insight through arts and theory as equal epistemological dimensions. New theories of playfulness come either from resources of art such as the *Situationists* (1959), or from contemporary alternate reality and urban game art theories. Scholars such as Salen

Image 7 *Blitz Play* playing cards, manual for a urban game, 2008. *Graphic* Jahrmann

(2005) identify today's world as a game space. In these new playful public spaces, play is a factor of an improved empowerment for agency. It allows posing questions about hidden ideologies inherent to present technologies of everyday life. Play principles can serve as a vehicle of practical intervention. As consequence of the observation of contemporary cultural techniques, it can be suggested that play no longer only stands for a symbolic action of large issues in life. Rather, the large issues in life have become play themselves, as a principle in economics and politics as well as in social interactions (Image 8).

Another art project developed together with Max Moswitzer is the arts server www.konsum.net. In this project, the operational ideas of a cooperative were gamified. The rules of play invited each user to contribute creatively to the technological network, the technical configuration, and the art content. The key element of this server configuration was that it was configured as GNU open source Slackware, a free software model allowing each member of the cooperative to get free access and domain space, which was at the time of the founding of Konsum in 1995 very rare and a political statement against the rising commercialization of the formerly free Web. To speak in the words of Geertz, the idea to configure a server

Image 8 Konsum server
logo, Jahrmann/Moswitzer,
1995. *Graphic* Moswitzer

as "play" that affects real life led to further social play with social structures, where play is an indicator for the social status of the player. This example demonstrates the application of play as motivation design through the combination of the different discourse systems of technology, game, play, and activism. The structure of such playful artwork points towards the term "gamification," listed in Google Trends[6] since 2010.

The concept of gamification is informed by a rhetoric of play, to use the term from the game scholar and designer Ian Bogost[7] (2011). In his book *How to Do Things with Videogames*, he describes how games are increasingly used for purposes other than entertainment. The experience of play often encourages the effect of a personal identification with an issue. Here the role of the player as an autonomous identity with particular possibilities of agency is crucial to the social ties and motivation achieved through game mechanics. Generally spoken, game mechanic principles are used in the concept of gamification in order to foster a strong motivation design for any context that is usually not a game. The experience facilitated in playful environments is exploited in gamification for business, marketing, and the organization of social processes. The concept is also used in a specific way in order to generate interest for contents in the cultural field, but also in health care and education. Designers, artists, museums, and advertising agencies use game mechanics as a tool to engage individuals as players. Therefore, contexts that are not games often profit from playful interaction. But these relations can be traced back to the time before computer games, even before electronic media, at the beginning of mechanization in the 1800s.

---

[6]Google Trends is a public Web facility, based on the Google search engine, that shows how often a particular search-term is entered relative to the total search-volume across various regions of the world and in various languages (https://www.google.com/trends/ [accessed: Jan 02. 2016]).

[7]Ian Bogost: Gamification is bullshit, Published online 2012. www.bogost.com/blog/gamification_is_bullshit.shtml [accessed: Jan 02. 2016].

3 Political Agency Through Play in History

Historic sources for political agency in relation to play and the direct empowerment of the individual through play can be observed in the case of the historic figure of "General Ludd." This iconic public play character appeared in the 1800s at the beginning of the Industrial Revolution in England. At that time, punch card-driven weaving machines were combined with steam power in the textile industry. The new mechanic looms dramatically changed conditions of labor and life of workers for the worse. The new steam-powered and punchcard-driven weaving machines became a symbol for the repression of workers and the vanishing of individual creation. Therefore, the mechanical loom became the target of destruction by weavers. Most surprisingly, in a situation of radical oppression and exploitation of the individual, not only brutal revolt but also playful strategies of protest were used. The following historic race against the machine demonstrates how media coverage supported the emergence of this political role-play character as ambiguous between reality and fiction in real political interventions. At that time, newspapers were still quite a new medium, but increasingly became relevant in public communication. Circular and causal feedback mechanisms between news items in the newspapers, and real action in the streets, encouraged the evolution of the fictional figure, General Ludd.

The media historian Sale (1999) describes the growing presence and distribution of print media technologies as the basis for the public use of the fictional character General Ludd by protesters against the mechanical weaving machines. According to newspaper reports of the time (Dec 20, 1811), the riots were led by an individual called Ludd. The *Nottingham Review* reported that "an apprentice named Ned Ludd smashed a master's machine near Leicester and hence gave his name to the action." From today's point of view it is more likely that a local Nottingham figure of speech was the source for the name of this character; "sent all of a lud" meant "struck all of a heap, or smashed" (Sale 1999: 2). Nevertheless, in newspaper reports Ludd had an increasing presence and was portrayed as an almost romantic figure, similar to Robin Hood. By repeated appearance and public enactment corresponding to the parallel contemporary media coverage of the figure of General Ludd, the character became a viable vehicle to be used for a new form of activist political role play. These aspects of the public reception of the activist as incorporation of an idea can be found again later on in the twentieth century, as described further on in this chapter, with gamelike characters in political activism, such as Luther Blisset.

Media coverage invites the creation of a fictional character. The historical "game" character of Ludd could not be punished or stopped because he never existed. For example, only as backlash to the media coverage, the protesters in historic England at the time of mechanization of work labor found courage for actions led by a number of self-proclaimed General Ludds. For them the joy of play, that normally role play would promise, was replaced by the necessities to survive without work and that led to the seriousness of real politics enacted in a role play. The effect of empowerment through role play remained a side effect of the invention of a leading character for anybody who wanted to become that leader for the duration of a political action, for the run against the machine. General Ludd became a placeholder for a community of empowered individuals who felt the need for a political leader of their protest. As far as it can be said from the angle of historical analysis, the figure of Ludd was not invented on purpose, but created spontaneously by a confluence of media reports and wishful thinking, and most important, the free will of the player to enact the role. The role of the leader was obviously taken by various people who only were individual workers before. This practice expressed the urgency for a political change but also for personal empowerment, which was achieved by an element of gamification of the process of political activity. Its aim was to protest against the technologies that destroyed the living conditions of workers. The role taken in such political agency was formed by the need to survive against control of the individual by mechanization. The existence of the fictional persona allowed the rioting weavers finally to protest against the technology that would destroy their living conditions.

4 Play Agency in Twentieth Century Political Theory

In a political role play many people are represented by one figure. Collective identities developed in such a playful practice are described in detail in political theories. The Italian Marxist Antonio Gramsci (1891–1937) tried to establish the necessity of artificial public figures, capable of articulating themselves within a wider social and political field.

The idea of a "collective intellectual" allowed bridging the practice of everyday life with an elaborate political consciousness. Most important, playful practice can be identified in this approach, although based on serious critical theory. Gramsci's writings had a great impact on a new form of public discourse in net cultures. He inspired thoughts about public figures of the twentieth century. In detail Antonio Gramsci speaks about the role of the artist as public intellectual, who has to design certain modes of interaction with a particular interest group and the general public. This function of a character is distinct from a natural persona or identity; it is highly comparable to what we know from the requirements of a character design in the discipline of game design. Usually an avatar[8] represents one player in a medium, but a game figure can also represent many identities in one character.

The application of playful mechanisms in society can be seen as gamified protest through role play. In that sense the phenomenon of role playing con-dividuals can be understood as precedents of a political game figure, which resurfaces in the playful forms of contemporary role play-based activism. A particular form of

---

[8]An *avatar* means, in a common sense of the networked age, a representation of a single self in a virtual play environment.

Image 9 Cover of the
original edition of *Handbuch
der Kommunikationsguerilla*,
1997. *Foto* Jahrmann

activist role play including a gamified form of media use was introduced by the
political arts practice of NEOISM.[9] Herein serious political protest was combined
with playful practices of role play, poetic writing under collective names, and the
subversive use of publishing media as well as irony (Image 9).

5 Luther Blissett: A Contemporary Activist Role Play

In 1997, the German *Handbuch der Kommunikationsguerilla* was published under
the names of Luther Blissett and Sonja Brünzels in the Verlag Rote Risse/Libertäre
Assoziationen, Berlin.

The book became known because of a copyright infringement of the German
do-it-yourself books titled, *Jetzt helfe ich mir selbst*, which can be translated as
self-empowerment. This incident already indicated one of the strategies described in
the book, the active play with media scandals and spectacles enabled only by and
through media. It draws on cultures of play, informed by pranks, hoaxes, and fakes,
but also includes role play as a vital element of protest against certain conditions of
life and work. The book collects methods for playful political interventions, which
can be classified as (media) spectacles, to protest the dominance of commercialization. For example, it describes how "adbusters" overwrite advertisements with
small creative interventions and totally change the meaning of the commercial
messages, and how to throw tarts on celebrities and political leaders at the right
"media" moment. Mainly methods of play for public interventions with the aim of
mass media coverage are introduced. With a focus on the play with media publicity

---

[9]Compare *"Anthologies of Neoism"* (Cantsin 1984: 4)

such practices were included in the collection of activist agency. The focus of the described activities was the "game mechanics" of playful protest.

As a kind of self-referentiality to the described practices the authors of the book were an anonymous collective of writers. The negation of individual authorship, the act of collective writing, can be understood as an activist statement in itself, which is playfully enacted through taking the role of Blissett. The role of Luther Blissett was exemplarily used as author for the handbook of playful political protest. In 1999, a novel entitled *Q* was published under the name of Blissett. It is about different aspects of conspiracy. A statement in relation to the book describes the role play enacted with the character of Luther Blissett:

> We have been active in the Luther Blissett Project since its beginnings, and integral parts of the Bologna scene since the late Eighties. We had and keep having problems with the Law. Our names are far from being important. Our biographies are even less relevant. We are the team that actually wrote "Q," and yet we are less than the 0.04% of the LBP. The fact that we are coming out does not comprise our self-spectacularization, we do not intend to give up our privacy to become (moderately) "young" fashionable novelists and talk show guests, which would be a very dishonourable end. If that ever happens we hope that other Blissetts will finish us off like wounded horses. Quite the contrary, our move is aimed at showing that we are a collective entity, not a single "Author." Behind Luther Blissett and behind "Q" as well there is no boss, no mysterious scholar, nor have we been the only Blissetts who contributed. It is the network the future of creative writing. (www.wumingfoundation.com [accessed: September 1. 2015])

In short, the name Luther Blisset is used by a number of individuals, in real life as well as online, for political interventions. Blisset appears as political activist; he holds speeches and publishes books. The use of this name by various authors and activists can be understood as political play, as a concept in capitalist societies for political action and agency (Image 10).

Image 10 El Sub Commandante Marcos. A political character with "game-props," the pipe and the mask, associated with a particular behavior. *Image Source* www.zocalo.com.mx/ seccion/articulo/ subcomandante-marcos-responde-a-criticas-1358032681 [accessed: Jan 5, 2016]

6 El Sub Commandante Marcos, a Political Play Under Revolutionary Conditions

Under critical political conditions, as in the case of the Zapatist movement in Chiapas, in moments and regions of political and social crisis the protection of the real identity of a person becomes a vital interest. Numerous media images of the political activist *El Sub Commandante Marcos*, the masked Zapatista leader in South America, demonstrate how the play type of "mimicry," as Caillois (1961) would express it according to his book, *Man, Play and Games*, can support political activism. The empowerment of certain groups of society, which are normally oppressed, is achieved through the identification with the character of El Sub. He gives a voice to those who normally cannot speak publicly in media. The condition to allow this public appearance is that the speaker is recognized as a political leader. This is in fact achieved by his appearance, his dramaturgical *"mis en scene,"* supported by costume elements, such as a black mask and a pipe. The leader is communicated as an indigene identification figure, a universal symbol for a tactic of deliberation by his statements and actions against suppression and colonial exploitation. Interestingly, through the identification of his identity through the mask, new role-play possibilities are opened up. The appearance of El Sub, his signature items of cap, mask, and pipe, allow the comparison of these costume elements available for everybody who wants to take that role, with costumes in live-action role-playing games. Photographs on the online chiapas indymedia, show female El Subs as well as male activists and children, all wearing the same costume indicating that they support or are associated with the political role play of the

Image 11 El Sub as role play, expressed in images of women and children posing in the costume of El Sub. *Image Source* https://revolutionarystrategicstudies.wordpress.com/2015/03/07/the-zapatista-womens-revolutionary-law [accessed: Jan 5, 2016]

character of El Sub,[10] including his political ideas, statements, and claims. Thus by mimicry everybody becomes El Sub. In the political fight, this masking of "faciality" is most relevant for self-empowerment (Image 11).

7 Conclusion: Towards a Ludic Society Through Activist Play

A tactical questioning of everyday life through play is expressed in various realms of political activism, contemporary game arts, and agency of the twentieth century. Its sources can be found in historic events at the beginning of mechanization. This perspective on play as intervention has been prepared by an anthropological analysis approach, which considered the contextual breadth of playfulness as a social strategy. Through different appearances and forms of play in society the quest for a political social utopia can be expressed in which play and game mechanics are used as political vehicles for agency. Social interactions are organized through game mechanics of voting, scoring, and competition. Social networking has become a matter of gamelike activities. In this context activist play serves as a societal tool for communication and intervention. Ideas of role play and performance converge in a new kind of critical play. Out of the entanglements of politics and play, of playful resistance, activist play demonstrates a trajectory from play practices to: play politics.

References

Bogost, I. (2011). *How to do things with videogames*. University of Minnesota Press.
Blissett, L., & Brünzels, S. (1997). *Handbuch der Kommunikationsguerilla*. Hamburg, Berlin: Rote Risse.
Caillois, R. (1961). *Man, play and games*. New York: Free Press of Glencoe.
Cantsin, M. (Ed.). (1984). *First manifesto of neoist performance and the performance of neoism*. New York: SMILE.
Geertz, C. (1973). Deep play: Notes on the Balinese Cockfight. In *The interpretation of cultures, theory of culture*. Cambridge, New York: The MIT Press.
Jahrmann, M. (2006). Ludics, a Nascent Art Research. In M. Santorineos (Ed.), *Gaming realities: A challenge for digital culture* (pp. 253–259). Fournos Athens.
Jahrmann, M. (2007). Ludic Society Manifesto. In A. Sudmann (Ed.), *Eludamos. 01/01. European Journal for Computer Game Culture*. Cambridge: MIT Gamebit Lab.
Jahrmann, M. (2008). Ludics. A new discipline in games studies. In *FROG. Future and Reality of Gaming, Conference Proceedings*, Vienna.
Jahrmann, M., & Moswitzer, M. (2006–2016).
Ludic Society, issue #1. (2005). *Ludics. New Bachelor Machines*, Graz, Madrid.
Ludic Society, issue #2. (2006). *Real Players. Ludic Interfaces*, Vienna, Zurich.

---

[10]http://chiapas.indymedia.org/display.php3?article_id=116929 [Accessed: April 12, 2009].

Ludic Society, issue #5. (2008). *Objects of Desire*, Bari/Vienna.
Ludic Society, VOID book. (2016). 10 year os Ludic Society—*100 Years of Dada*, Zürich.
Ludovico, A. (2008). *Neural* Issue #30, Dangerous Games, Bari Italy.
Salen, K. (1999). The achievements of General Ludd. A brief history of the Luddites. In *The Ecologist* (Vol. 29), Ecosystems Ltd., London.

Author Biography

Margarete Jahrmann is an artist, curator, and researcher in activism, games, and arts with an international record of exhibitions and conferences. She has received major media arts awards, such as the Prix Ars Electronica 2003 and Transmediale Berlin 2004. Since 2006 she has been professor for game design at the University of Arts Zürich and since 2000 senior lecturer of art and technology at the University of Applied Arts in Vienna. In 2006 she founded the Ludic Society in Plymouth, United Kingdom and since then edits the Ludic Society magazine (ludic-society.net). In 2013 she created an alternate reality exhibition at Kunsthalle Vienna and edited a corresponding book together with Brigitte Felderer under the title *Play & Prosume. Schleichender Kommerz und schnelle Avantgarde* (Verlag für Moderne Kunst Nürnberg). In 2015 she worked and published "Agon & Ares, War and Games," an article published in 2016 in the book *Agon & Ares* (edited by Ernst Strouhal, Springer Wien/New York). In 2015 she also published "Deep Play. Arts Experiments as Strategies of Participative Research" in the *Journal for Research Cultures. Strategien künstlerischen und wissenschaftlichen Arbeitens* (Issue #1, edited by Matthias Tarasiewicz, Research Institute for Arts & Technology, RIAT). In 2016 she edited the *VOID Book* (Verlag Ludic Society Zurich) and presented it at Cabaret Voltaire Zurich at the occasion of 100 years of DADA.

Part IV
Place, Reality, Meaning

Tricksters, Games and Transformation

Maggie Buxton

Abstract This chapter weaves together trickster figures, emerging game formats and transformative learning theories, and explores relationships between these discourses and experiences. Tricksters are argued to have new twenty-first century Western relevance. Augmented Reality and Alternate Reality Games are described as trickster tools. A potential relationship between these tools and transformative learning is outlined.

1 Introduction

I want to introduce you to tricksters, those mysterious and mischievous figures that pop up in the narratives of cultures across the world.[1] Ananse, Èṣù, Reynard, Hanuman, Maui, Hermes—you may have heard childhood tales featuring the names of these demi-gods, gods, humans, heroes or heroines, animals and insects. They been variously described as 'archetype, myth, and life symbol' (Lundquist 1991); a culture hero (Luomala 1949); and an interstitial, hybrid being that is 'neither a god nor a man, neither human nor animal' (Pelton 1993, p. 137).

This chapter brings together trickster figures, emerging game formats and transformative learning theories. I establish relationships between these discourses and experiences and propose that through examining them together each may be seen in a new light. Indeed, Lewis Hyde (1998) puts forward the idea that placing trickster stories next to 'imagination in action' allows each to illuminate the other. I class the game formats described in this chapter as 'imagination in action' in its truest sense. I begin by looking at tricksters including an argument for their twenty-first century Western relevance; describe how certain game formats may be

---

[1]Some aspects of this article were published already as part of the authors Ph.D. thesis noted in reference list

M. Buxton (✉)
AWHIWORLD, 2 St Marys Place, Onerahi, Whangarei, New Zealand
e-mail: maggiebuxton@gmail.com

© Springer Nature Singapore Pte Ltd. 2018
D. Cermak-Sassenrath (ed.), *Playful Disruption of Digital Media*,
Gaming Media and Social Effects, https://doi.org/10.1007/978-981-10-1891-6_14

seen as trickster tools; and posit a potential relationship between these tools and transformative learning and identify a need for further research in this area.

2 Trickster

Certain traits are common across trickster stories. I summarise a few such traits but guided by Hynes and Doty (1993), I aim to steer a course between 'blunt universalism and a relativist view which maintains that tricksters are so culturally specific that they cannot deliver universal messages' (p. 2). Tricksters are creatures that turn in many different ways, shapeshifting their way into and out of tricky situations. They are as 'shifty as an octopus, colouring themselves to fit in surroundings, putting on a fresh face for each man or woman they meet' (Doty 1993, p. 53).

Tricksters are creatures of the margins, and the spirit of the in-between. As Hyde (1998) notes, 'every group has its edge, its sense of in and out and trickster is always there... Trickster is neither god of the door leading out or the door leading in – he is the god of the hinge' (p. 9). Tricksters inhabit interstitial places, situations and experiences, seizing the opportunities that pass by.

Tricksters travel across multiple realities and into parallel worlds. Trickster fables illustrate a way of thinking that is multidimensional. Creatures wander from place to place talking to waterfalls, gods and beetles, fornicating with animals and dignitaries, and opening portals to other realms peopled by magical creatures. They play magical tricks with perspective, matter and physics and with reality itself. Tricksters blur the distinctions between the natural, supernatural and un-natural: The Winnebago Trickster, for example, piles his enormously long penis on his back, sending it out to fornicate on far shores; Anansi climbs a ladder to the Gods with a pot full of wisdom on her back; Maui snares the sun and hooks up entire landmasses.

Truth and fiction are often blurred in trickster fables. According to Hyde (1998), when tricksters lie and steal it is usually to disrupt dichotomies of truth and property and therefore create new possibilities. Trickster speech confuses the distinctions between lying and truth-telling, undercutting fictions by which reality is shaped (p. 213). In one Èṣù tale, the trickster makes a hat that has one side black and one white in order to stir trouble between friends. Each friend swears that the hat is either black or white depending on what side of the field they stood when Èṣù rode through on his horse. Finally, they realise they had been tricked and that each had been telling the truth all along. In this tale, one of the underlying messages relates to the multiplicity of perspective.

Tricksters facilitate connections between humans, and connections between humans and gods (Doty 1993, p. 51). The trickster god Hermes organises the social cosmos, working out the interconnections among people, boundaries between nations, and realignments of military of political power (Doty 1993, p. 56). According to Hyde (1998), cultures need figures whose purpose is to reveal and disrupt the very things that cultures are based on—that is how cultures maintain their liveliness and durability. In fact, Hyde (1998) describes tricksters as 'artus workers', noting that they work at the joints of society, creating separation, setting others at odds or placing boundary markers in new and unusual places. They make or remake the articulated world.

Trickster stories often pivot around key points of transition in life—the transition between life and death being an essential one. It is at these points that transformation of self occurs (Combs and Holland 1996, p. 84). A trickster, then, 'symbolises humankind's self- transcending mind and the quest for knowledge and the power that knowledge brings' (Ricketts 1993, p. 87). They are therefore symbolic of, and catalysts for, reflective consciousness.

Finally, tricksters embody the spirit of play. It is their mischievous way of engaging with the world, their ability to turn traditionally serious topics (such as death, identity and power) into amusing twists of fate and luck, and their way of making us laugh at them and ourselves is the essence of trickster spirit. Davis (2010) discusses the games tricksters play with perspective, and notes that 'trickster shrines are placed at crossroads and at the market—trickster is a master of the crossed purposes that define networks of exchange and circuits of desire'. Hyde (1998) points out that these figures play with cultural webs, the underlying patterns that make up how we are as a society. This includes subverting the boundaries of how we know species and gender. As Doty (1993) notes, the trickster clearly distinguishes itself 'by its frequent association with shape shifting and situation inversion … not even the boundaries of species or sexuality are safe, for they can be readily dissolved by the trickster's disguises and transmorphisms' (p. 36).

So trickster is a shapeshifter, a traveller across worlds, a challenger and reformer of culture, a figure that inspires, if not exemplifies, a coming to knowing in the world, and trickster does this through different forms of play. Trickster is, in essence, a transformative figure.

3 Trickster Today

Life in the second decade of the twenty-first century is multiplicitous. According to Spretnak (1997, p. 1): 'The disintegration in recent years of so much that previously seemed stable is disconcerting to anyone who has been paying attention'. Foucault (1986) views 'modern' society to be in an 'epoch of simultaneity, an epoch of juxtaposition, the epoch of the near and far, of the side-by-side, of the dispersed'. He describes a shift from linear experience of space and time to one that is multiplicitous in nature. Levinson (1999) also refers to 'multiplicity' as the spirit of the digital age, a new type of world where traditional arts subjects are becoming obsolescent, and traditional academia is undermined, through the World Wide Web's creation of a malleable curriculum for living. According to Relph (2007) there has been a 'clear move away move away from the objective, rationalist perspective … to a view that acknowledges the validity of many different perspectives'.

Davis (1998) also sees the Web in particular as inherently multiplicitous with its lack of central power, proliferation of media forms and the lateral links it creates between 'various networks, autonomous programs, and genres of expression'. Arthur (2006) uses the word 'plethora' to describe the new forms of conceiving, organising and articulating knowledge that the Web has created. It is within this context of connection, disruption and multiplicity that I believe the trickster figures find new relevance. Indeed, as Hynes (1993) notes:

> Trickster pulverizes the univocal and symbolizes the multi-valence of life. Embodying this multi-vocality, the trickster himself eludes uni-vocality by escaping from any restrictive definition: the trickster is always more than can be glimpsed in any one place or in any one embodiment. … The trickster disorders and disassembles.

Tricksters thrive in situations where real and unreal are no longer defined categories. I believe Augmented Reality (AR) and Alternate Reality Games (ARGs) exemplify the spirit of the trickster today.

4 Twenty-first Century Play

Mixed reality (MR) games include game formats that mix the physical, digital and imaginal worlds. Two key game formats are AR games and ARGs. The term AR refers to applications that allow digital image, audio or text to 'augment' our day-to-day experiences. This can happen in a variety of ways, but usually through the camera view of one's GPS-enabled mobile device (phone, tablet or glasses). It is the sense of reality achieved by superimposing virtual objects and cues upon the material world in real time (Carmigniani and Furht 2011) either directly through a viewing window ('look through' view) or indirectly (e.g. via video) i.e. 'heads down' view.

Some authors take exception to the idea that AR automatically *enhances* surroundings the more general term MR instead (Champion 2011). Augmented (mixed or hybrid) Reality mobile games utilise mobile phones with high-quality sensing, sound and imaging systems combined with AR applications. Using this technology, educationalists, artists, and community activists are experimenting with creating experiences that allow a rich interchange of data both from user to facilitator, but also collaboratively from user to user, and user to the environment (Klopfer 2008).

Early examples of these games (used in educational settings) were: *Environmental Detectives* (Klopfer 2008; Klopfer and Squire 2007) and *Mad City Mystery* (Squire 2007) both of which involved some sort of mystery to be solved. While these serious games are still evolving, more recently, AR mass entertainment games have emerged such as *Ingress* (Niantic Labs/Google), *Clandestine Anomaly* (Zenfri Inc.), Pokemon Go (Niantic/Nintendo), and my own organisation's (AwhiWorld Ltd.) location-based MR game-like experiences.

ARGs are a narrative-based collaborative game form that traverses a mix of digital and physical media (Martin and Chatfield 2006). They are another emerging educational tool with prototype games and experiences emerging within mainstream educational settings (Buxton 2008; Colvert 2009; Gislén et al. 2007). Increasingly, as in the case of *Ingress*, they use locative mobile and AR to create games that cross boundaries of materiality, place and space. Experiences like these allow players actively co-create and collaborate in the storytelling process than that simply view from a distance.

These MR game formats are combining to create game-based learning experiences. AR mobile phone applications allow us to view a world that is hidden from everyday view. ARGs blur the line between day-to-day reality and imaginary play. Essentially, these types of experiences disrupt and expand what we understand to be real.

It is only through the ubiquitous adaption of mobile devices that these game forms are possible. Mobile and wearable devices allow access to other worlds: fantasy landscapes; cultural realities (ethnic, political, religious, moral, etc.) and individual life-worlds (through comprehensive personal social media sites). Users can fly up into the sky and back down again, by tokens and souvenirs, smuggle money, connect with loved ones, all without any official border checks. Conversations can occur with people thousands of miles away and ancestors can speak to us while we stand on their grave—the dead resurrected in digital form.

Through apps, the device shapeshifts into an object that facilitates interaction scientifically, educationally, socially, financially, culturally, and in the work of the leisure. Users are able to play a number of roles—and in the locations that they choose.

5 Trickster Tools

I see the game formats described above as 'trickster tools'. Tricksters are often hybrid creatures that evade categorisation, and so do these game formats. Mobile devices are hybrids of a number of technologies, and as mentioned, can shapeshift (and allow us to shapeshift) at will.

Tricksters move between realities. In indigenous societies, shamans travel and communicate interdimensionally using mirrors, sacred objects and plants (MacDonald et al. 1989). Hand-held devices such as mobile phones and tablets are also portals, allowing us to not only communicate with other worlds and realities, but also time travel between past and future.

Tricksters often disrupt order to illuminate and play tricks and games in order to educate. In these games, formats surprise and mysteries are often a key part of the game process. As Silva and Delacruz (2006) note, even the uncertainty of playing a game in a 'real-world' space (instead of the relative 'safety' of a classroom) can create unexpected surprises as non-participants and the natural environment inadvertently insert themselves in the game process.

6   Transformative Learning

Transformative learning is an emancipatory educational philosophy that is concerned with the catalysts and conditions that expand the consciousness of adults and release them from unexamined, unconscious belief systems. Disorientation and disruption are included in the range of conditions that need to be in place for sustainable transformation to occur. Mezirow (2000) discusses a range of disorienting phenomena, from small irritations to dilemmas such as death, illness, divorce, material loss and betrayal.

Transformative learning involves 'transforming taken-for-granted frames of reference (meaning perspectives, habits of mind, mind sets) to make them inclusive, discriminating, open, emotionally capable of change, and reflective' (Mezirow 2000, p. 8). Transformation often occurs when something unexpected happens (Cranton 2006) When an individual encounters a situation that does not fit in with assumptions and expectations of how things should be they either reject the situation or question their assumptions and expand their frame of reference to include this new point of view. 'When people critically examine their habitual expectations, revise them, and act on the revised point of view, transformative learning occurs' (Cranton 2006, p. 19). Other researchers have also shown that profound transformation is often associated with intense suffering or crisis. Difficult life events shatter defences and leave us vulnerable and open—this creates the conditions for new perspectives to emerge (Schlitz et al. 2007).

I propose that trickster stories, as examples of disruption and disorientation, and movement between states of unknowing and knowing, are fables for transformative learning. Tricksters, due to their hinge-dwelling existence and boundary crossing antics, continually challenge humans to examine what is on the outside with what is on the inside (Pelton 1980, p. 234). Through this reflective process, they shift the patterns that create individuals and societies, they are therefore described by Hyde (1998, p. 257) as 'second order articulators'.

Hermes' actions allow humans to move to a higher level of awareness or insight (Doty 1993, p. 55). Maui represents the inner tugs between 'me' and 'we' and is a kind of divine scapegoat where Oceanic peoples can escape rigid protocols between the sacred and profane (Luomala 1949, p. 29), and in so doing can reflect on these societal rules from a distance. Simply by existing between the bounds of conventions tricksters undermine underpinning assumptions and frameworks. They re-articulate our ways of knowing the world. Trickster tools and practices confound categories of materiality and force us to cognitively engage in increasingly sophisticated ways.

7 Disorienting and Disrupting Playful Technologies

Transformative education professionals have usually created disorienting dilemmas through traditional methods such as case study analysis, group work, story-sharing, field trips, conscious reflection and meta-theoretical study. These methods aim to disrupt taken-for-granted assumptions about many different areas including, at the deepest level, fundamental epistemological and ontological beliefs. The two-game formats discussed earlier in this chapter provide an opportunity to disorientate and disrupt—and do this playfully rather than painfully.

According to de Souza and Silva (2006), these MR games 'allow an imaginary playful layer to be overlaid on physical landscapes, changing players' perception of spaces and merging borders between the game, and spaces usually associated with day-to-day life'. By disturbing fundamental assumptions about what constitutes reality, and expanding ontological frames, these games may foster epistemological flexibility. This perhaps furthers the work of Turkle and Papert (1990), who, in discussing the use of computers in the classroom, argue certain technological developments (such as artificial intelligence, programming philosophy) create an opening for 'epistemological pluralism'.

As an example, Èṣù used a differently coloured hat to confuse two friends who had conflicting stories. In a game format, via the portal of a phone, conflicting narratives are juxtaposed forcing the audience to step back and reflect on their assumptions about facts and truth. By augmenting a location with stories, the categories of 'alive' and 'dead', 'real' and 'unreal' are blurred, forcing us to make sense of communities in a new way. In a way, the fundamental nature of these game forms is disruptive. By playing them, is it possible to expand our cognitive capacity and transform or shift our consciousness?

It could be argued that rather than disrupting assumptions about how learning should happen, AR and ARG game formats are actually just preparing us to be good consumers of online content in a neoliberal marketplace—fodder for the entertainment industry. However, there is room to imagine that the multiplicity of our world can be understood through a reflective consciousness, and that this can be fostered through the trickster tools available to us in the twenty-first century.

8 Conclusion

The archetypal figure of the trickster has a number of traits that relate to Augmented and Alternate Reality game forms. In these games, it is possible to travel to other worlds and to shapeshift into other forms. The disruptive, disorienting effect of these forms of play is potentially forms of transformative learning. It appears possible to see the trickster not just as a figure from the past, but a figure relevant to the present—a figure that, among many interesting traits, fosters reflective consciousness through playful tricks. In an age where it seems critical to challenge assumptions about justice, parity and sustainability—perhaps lessons from the trickster are even more relevant now than ever before.

References

Arthur, P. L. (2006). Hypermedia history: Changing technologies of representation for recording and portraying the past. *InterCulture: Revisioning History, 3*(3), 1–25. Special issue

Buxton, M. (2015). *Tricksters, technology and spirit: Practicing place in Aotearoa-New Zealand.* AUT University, Auckland. Retrieved from http://aut.researchgateway.ac.nz/bitstream/handle/10292/8847/Final%20Proof%20June%202015.pdf?sequence=1.

Buxton, M. (2008). *Grig: Tricksters, unfiction and alternate realities.* Retrieved from http://nadine.be/ARCHIVES/old/tricksters-unction-alternate-realities-maggie-buxton.

Carmigniani, J., & Furht, B. (2011). Augmented reality: An overview. In B. Furht (Ed.), Handbook of augmented reality (pp. 3–46). New York, NY: Springer. https://doi.org/10.1007/978-1-4614-0064-6_1.

Champion, E. (2011). Augmenting the present with the past. In *Playing with the past.* London, UK: Springer. https://doi.org/10.1007/978-1-84996-501-9_7.

Colvert, A. (2009). *Peer puppeteers: Alternate reality gaming in primary school settings.* Presented at the meeting of the DiGRA: Breaking New Ground: Innovation in Games, Play, Practice and Theory, West London, UK.

Combs, A., & Holland, M. (1996). *Synchronicity: Through the eyes of science, myth, and the trickster.* New York, NY: Marlowe and Company.

Cranton, P. (2006). *Understanding and promoting transformative learning: A guide for educators of adults.* San Francisco, CA: Jossey-Bass.

Davis, E. (1998). *Techgnosis: Myth, magic and mysticism in the age of information.* London, UK: Serpents Tail.

Davis, E. (2010). *Nomad codes: Adventures in modern esoterica.* Portland, OR: Yeti/Verse Chorus Press.

de Souza e Silva, A. (2006). From cyber to hybrid: Mobile technologies as interfaces of hybrid spaces. *Space and Culture, 9*(3), 261–278. Retrieved from http://sac.sagepub.com/content/9/3/261.

de Souza e Silva, A., & Delacruz, G. C. (2006). Hybrid reality games reframed: Potential uses in educational contexts. *Games and Culture, 1*(3), 231–251. Retrieved from https://doi.org/10.1177/1555412006290443.

Doty, W. G. (1993). A lifetime of trouble-making: Hermes as trickster. In W. J. Hynes & W. G. Doty (Eds.), *Mythical trickster figures: Contours, contexts and criticisms* (pp. 46–65). Tuscaloosa, AL: University of Alabama Press.

Foucault, M. (1986). Of other places. *Diacritics, 16*(1), 22–27.

Gislén, Y., Löwgren, J., & Myrestam, U. (2007). Avatopia: A cross-media community for societal action. *Personal and Ubiquitous Computing, 12*(4), 289–297. https://doi.org/10.1007/s00779-007-0152-5.

Hyde, L. (1998). *Trickster makes this world: Mischief, myth and art.* New York, NY: North Point Press.

Hynes, W. J. (1993). Mapping the characteristics of the trickster: A heuristic guide. In W. J. Hynes & W. G. Doty (Eds.), *Mythical trickster figures: Contours, contexts and criticisms* (pp. 33–45). Tuscaloosa, AL: University of Alabama Press.

Hynes, W. J., & Doty, W. G. (1993). Introducing the fascinating and perplexing trickster figure. In W. J. Hynes & W. G. Doty (Eds.), *Mythical trickster figures: Contours, contexts and criticisms* (pp. 1–12). Tuscaloosa, AL: University of Alabama Press.

Klopfer, E. (2008). *Augmented learning: Research and design of mobile educational games.* Cambridge, MA: The MIT Press.

Klopfer, E., & Squire, K. (2007). Environmental detectives—the development of an augmented reality platform for environmental simulations. *Educational Technology Research and Development, 56*(2), 203–228. https://doi.org/10.1007/s11423-007-9037-6.

Levinson, P. (1999). *Digital McLuhan: A guide to the information age.* London, UK: Routledge.

Lundquist, S. E. (Ed.). (1991). *The trickster: A transformation archetype* (Vol. 11). San Francisco, CA: Mellen Research University Press.

Luomala, K. (1949). *Māui-Of-A-Thousand-Tricks: His Oceanic and European biographers.* Honolulu, HI: Bernice P. Bishop Museum.

MacDonald, G. F., Cove, J. L., Laughlin, C. D., & McManus, J. (1989). Mirrors, portals and multiple realities. *Zygon, 24*(1), 39–61.

Martin, A., & Chatfield, T. (2006). IGDA ARG SIG Whitepaper. Retrieved from http://wiki.igda.org/index.php/Alternate_Reality_Games_SIG/Whitepaper.

Mezirow, J. (Ed.). (2000). *Learning as transformation: Critical perspectives on a theory in progress.* San Francisco, CA: Jossey-Bass.

Pelton, R. D. (1980). *The trickster in West Africa: A study of mythic irony and sacred delight.* Berkeley, CA: University of California Press.

Pelton, R. D. (1993). West African tricksters: Web of purpose, dance of delight. In W. J. Hynes & W. G. Doty (Eds.), *Mythical trickster figures: Contours, contexts and criticisms* (pp. 122–140). Tuscaloosa, AL: University of Alabama Press.

Relph, E. (2007). Spirit of place and sense of place in virtual realities. *Techné: Research in Philosophy and Technology, 10*(3), 17–25. Retrieved from http://scholar.lib.vt.edu/ejournals/SPT/v10n3/relph.html.

Ricketts, M. L. (1993). The shaman and the trickster. In W. J. Hynes & W. G. Doty (Eds.), *Mythical trickster figures: Contours, contexts and criticisms* (pp. 87–105). Tuscaloosa, AL: University of Alabama Press.

Schlitz, M. M., Vieten, C., & Amorok, T. (2007). *Living deeply: The art and science of transformation in everyday life.* Oakland, CA: New Harbinger Productions Inc.

Spretnak, C. (1997). *The resurgence of the real: Body, nature, and place in a hypermodern world.* Reading, MA: Addison-Wesley Publishing, Inc.

Squire, K. (2007). Mad city mystery: Developing scientific argumentation skills with a place-based augmented reality game on handheld computers. *Journal of Science Education and Technology, 16*(1).

Turkle, S., & Papert, S. (1990). Epistemological pluralism: Styles and voices within the computer culture. *Signs, 16*(1).

Author Biography

Maggie Buxton is an accomplished producer, educator, consultant, and place innovator. She is the founder and director of the social enterprise AwhiWorld.

Maggie has extensive international experience in facilitation, strategic development, and communications consulting. This includes working with eco-village communities in West Africa, Latin America, and Northeast Scotland, social enterprises in the United Kingdom, experimental designers and artists in Brussels, climate change initiatives in Beijing and Singapore, and large political institutions and corporations in Europe.

The last 15 years of her career have been spent exploring and experimenting with creative and innovative methods, tools, and technologies including alternate reality gaming (ARG, and other transmedia formats); augmented reality; locative media; and locative mobile applications. Her work is carried out in partnership with indigenous organizations (iwi/tribal groups), senior/elder institutions, migrant groups, maker groups, activists, entrepreneurs, and experimental artists. Her work is transdisciplinary, fundamentally connected with place and designed to weave different worlds together or create new worlds collaboratively. She is currently working with the concept of place as an emerging media channel and site for facilitating, generating, and investigating parallel realities.

Makin' Cake—Provocation, Self-confrontation, and the Opacity of Play

Daniel Cermak-Sassenrath

Abstract Players can do the most extraordinary things in games without raising an eyebrow. Here, three specific questions are discussed: What do objects and actions in games mean? How are these meanings constructed? By and for whom? It is argued that players most naturally understand and know perfectly well what their actions in games mean and how they relate to everyday life: Actions in play are blank, and mean nothing. Meaning is only created within play, in a fluid, dynamic, and collaborative process, over time, based on an implicit understanding and shared practice. Meaning is not seen as abstract truth and values, for all times and across all cultures, but relative: something *gains* meaning for somebody, in a particular situation and context. The interactive installation *Makin' Cake* demonstrates the issue of the meaning of play activities within and without play by providing an immediate and provocative experience to players and spectators.

1 Introduction

Players can do the most extraordinary (e.g., physical or violent) things in games without raising an eyebrow. Here, three specific questions are discussed: Which meaning do actions and objects have in play? How do they get it? For and from whom?

There are many games in which players are asked to perform quite dodgy tasks that are essential to the game play, and without which the games do not proceed. Stealing an object from each other (as in many ball games), physically assaulting, potentially hurting and injuring each other (as in many contact sports), or taking off each other's people (as in chess) are examples from traditional games. In digital games, digging up a human bone from a graveyard in *Monkey Island 2* (1991)

D. Cermak-Sassenrath (✉)
Center for Computer Games Research, IT University of Copenhagen,
Copenhagen, Denmark
e-mail: dace@itu.dk

© Springer Nature Singapore Pte Ltd. 2018 223
D. Cermak-Sassenrath (ed.), *Playful Disruption of Digital Media*,
Gaming Media and Social Effects, https://doi.org/10.1007/978-981-10-1891-6_15

appears relatively harmless compared to activities in some of today's games (e.g., the (optional) massacre of unarmed civilians at an airport in *Call of Duty: Modern Warfare 2* (2009) (Anonymous 2009) and the killing of a large portion of the population of the city of Stratholme in *Warcraft 3: Reign of Chaos* (2002) (WoWWiki 2013)). Although occasionally there is some discussion (e.g., on the baptism sequence in *Bioshock Infinite* (2013) (Polygon 2013; Volta 2013; Irrational Games Forums 2013)), it appears not to be a major problem for the majority of players to carry out these activities (cf. WuShogun212 2013), neither in traditional games nor in digital games.

It can be argued that experienced players of popular games are desensitized zombies or social freaks who are unaware and ignorant of their actions, or that they have become nerdy media geeks who are able to decode games from an analytical distance without being bothered by their content. Here, an explanation is offered that argues that players most naturally understand and know perfectly well what their actions in games mean and how they relate to everyday life.

The discussion in this chapter uses the participatory installation *Makin' Cake* as an example. It demonstrates a conflict between play and the everyday world and provides participants with an immediate and challenging first-hand experience and a "moment of self-confrontation" (Murray 1997, 54). The game confronts people and makes people confront "an authentic but disquieting side of [themselves]" (Ibid.). The setup as an installation (a 1950s kitchen), where playing the game also means performing before an audience, invites critical reflection and discussion.

Makin' Cake is a skill-based competitive single- player game with a twist. Players type-in 1950s cake recipes as fast as they can and get points for every word; for instance, "put flour and sugar into bowl," "mix flour with sugar," and "add two eggs." The twist are extra points for swearing. *Cake* grants players a substantial advantage in achieving a high score, when and if they swear. Nobody is obliged to swear, and the game is playable without it; but top scores are then difficult to achieve. The game that people can play is innocent, with several levels, recipes, and high-score lists. The game becomes vulgar as soon as players find out that swearing produces extra points. And it is scary how rapidly and how willingly players accept the boundaries and rules imposed by the medium.

2   The *Makin' Cake* Installation

Makin' Cake (Fig. 1) is an interactive installation[1]; participants can play a single-player competitive game. To start the game, players select a recipe from the main menu (Fig. 2).

---

[1]The installation has been shown at the Creativity & Cognition (C&C) 2013 conference in Sydney, Australia; at the Design and Semantics of Form and Movement (DeSForM) 2013 conference in Wuxi, China; and at The 9th Australasian Conference on Interactive Entertainment (IE) 2013 in Melbourne, Australia.

Fig. 1 Title screen (screen shot)

Fig. 2 Main menu (screen shot)

Fig. 3 In-game (screen shot)

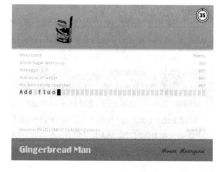

Each recipe is divided into nine steps. The ingredients to be used and the actions to be performed in each step are displayed at the top of a screen/projection (Fig. 3). An image of a packet of baking flour means that in this step, flour needs to be added to the dough, and so on. In each step all the necessary ingredients and actions (up to three) have to be typed-in. Players press return/enter to go to the next line and the next step. For each line they receive a score calculated from the number and length of the words they entered.

The text input (see Table 1) is checked using three separate word lists: certain keywords must occur in a step, e.g., "flour," "bowl," and "stir"; if the words do not occur the player cannot proceed to the next line or step of the recipe. The check of

Table 1 Player recipe for a
Gingerbread Man
(C&C 2013)

Player recipe 2: Gingerbread Man
Anonymous Chef
Sugar and golden syrup bitch
Eggs is what i love
Water for me daughter
Mix for the chicks
Flour power
Vanilla pepper ginger
Powder
Mix
Bake

keywords is performed quite leniently; otherwise it would be a tedious game with players spending their time guessing what the images represent. If, for example, the recipe asks for "baking powder" only "baking," "powder," "bak," or "pwd" is required; "butter" can be substituted for "shortening" or "margarine," and there is a whole range of keywords accepted to "mix" the dough; words are accepted in singular or plural. A second list is used to recognize general English words, and a third list to identify foul language. Long words give exponentially more points than short ones.

Points are only awarded for words that are not required keywords: Every letter gives 1 point multiplied by word length (e.g., "cat" gives $3 \times 3 = 9$ points, i.e., the score for one word is calculated by $wordlength^2$); letters of swear words give 10 points (e.g., "bloody" gives $6 \times 10 \times 6 = 360$ points; i.e., $cursewordlength^2 \times 10$). Words that are used repeatedly give fewer points each time until they do not give any points at all. Numbers and one-letter words are not counted.

The nine recipes have different time limits. Recipe 1 (One Egg Cake), for instance, has a time limit of 90 s, level 2 (Gingerbread Man) one of 80 s. During each level a kitchen timer indicates the time running out. If the time is up in the middle of typing-in the recipe, the game is over. Leftover seconds do not translate into points. To avoid players typing-in four-letter words in the last line of the recipe in case they have time left, every line is limited to 40 letters.

After the round players are informed if their score was high enough to be invited to join the champions' hall of fame. In this case, they can enter their name and have their picture taken. The photo high- score list adds a second layer to the game, that is, how and if players stand by their recipes full of swearing; also, players can play off the props available in the kitchen setup (e.g., large spoon, chef's jacket, apron, hat; Figs. 4 and 5). The installation setup comprises a large display, so spectators can see the typing (Fig. 6). The top recipes are displayed while nobody is playing (Fig. 7).

The recipes used in the game are based on actual cake recipes. The recipes for the "One Egg Cake by Mona Lemmon," "My Buttermilk Cake from Hilda Earhart's Recipe Box" (dated 1958), "Pineapple Souffle from Betty Arledge," "Cranberry Nut

Fig. 4 Setup on a kitchen
counter look-alike
(C&C 2013)

Fig. 5 Player posing with
props for a photo for the high
score list (C&C 2013)

Fig. 6 Player and spectators
(DeSForM 2013)

Cake" by Mrs. Alley, "Prune Velvet Cake by Eva Wiedrick," and "Lazy Daisy Oat-meal Cake" by Margie Frump are genuinely vintage (PenVampyre@aol.com 2011). The collector found the recipes "from the 1950s and 1960s"

> on handwritten recipe cards cached in colorful old tin boxes, scribbled in the margins of ads or on bits of paper slipped in between the pages of well-worn cookbooks gathered from flea markets, auctions, estate sales, and any place that treasured old recipes are wont to hide (Ibid.).

The recipe for the Apple Crumble cake is from the 1940s (Ekins 2009), the one for the "Fruit Pavlova by Mrs. Bernadette Haddock, of 'Dandaraga,' Tara, Western Downs" from 1954 (Anonymous 1954, 11).

Fig. 7 High-score list entry
with recipe and player photo
(C&C 2013)

Few games use swearing as a game mechanic, for instance, *Bleep* (jmr24 n.d.).
Bleep is a *Flash* game in which players enter swear words with a time limit. There is
no visual output; only sound and the changing score acknowledge successful swearing. In the online high-score list players' names, country flags, and number of successfully entered words are displayed, but not the words themselves. The top player
apparently entered 116 words in 60 s. The game is described on one website[2] as a
"[t]est" of the player's "vile knowledge of the ancient and mystic art of swearing,"
and is categorized as an educational game.

In several popular games swearing is included as part of their setting, as sound
effects in-game or in cut scenes. In his article, "10 Most Swearing-est Games"
(McNeilly 2008), Joe McNeilly lists *Beat Down: Fists of Vengeance* (14 swears),
GTA San Andreas (20), *25 to Life* (24), *Saints Row* (32), *50 Cent: Bulletproof*
(48), and *Scarface* (70) among others. *Scarface* apparently has a "swear-on-demand
option" (Ibid.). Other games criticized (rather than praised) for swearing are *Far Cry
3*, *Bulletstorm* (Laura 2013), and *The Walking Dead* (2013); but according to Laura
(2013), Lara Croft gets it right "in the new Tomb Raider reboot."

3 Media Are Blank

Media are appropriated in context and through a process. Meaning is not embedded
or hidden in medial texts, and is not only uncovered by their readers, but is creatively
produced and (dis-)agreed upon, potentially anew in every use, depending on the
needs, wishes, and expectations of the readers, in an active, dynamic, collaborative,
and creative process. This is a shared process that is neither fully determined nor
arbitrary (Hall 1993); people are able to agree on a meaning with enough overlap to
achieve (a considerable degree of) stability and usability. For every person, the creation of meaning is an individual, subjective, and inherently embodied process. The
process of the construction of meaning is not abstract, distant, or based on observation, but is based on becoming and being a participant in it. Something needs to be

---

[2]www.fupa.com/play/Education-free-games/bleep.html (May 14, 2013).

connected by and with a person to have meaning for that person: If the person is not present, nothing has meaning for him or her. Media use is active, and takes different forms depending on the particular medium. Media are different in specifically this aspect: the possibilities they offer for user participation. This is not limited to interactive media, and media users do not need to be able to change a medium's course and the outcome. Everything media users experience is produced by them themselves. In games, the meaning of play actions is created through a tight coupling between action and consequence.

3.1 Blank Texts

Media do not contain fixed messages that (only) need to be uncovered, recognized, or identified (see Winter 1995, 110–1), but meaning is only constituted or produced by media users, depending on their expectations, situations, and needs, for example, in specific contexts. Relevance appears to influence the process more than preference (Fiske 1988 in Winter 1995, 108). The author of a medial text does not define its meaning, but its readers and their use of it (Winter 1995, 222). Media use is an active, creative, on-going, and fluent process between audience and artist, or rather, between audience and work. The song "Used to Love Her" (Guns n'Roses 1988) advises the listener to "take it for what it is," but it is not obvious what the something is.

> Active audience theory also confirms the presence of polysemic content, which challenges the view that media texts are closed, or can be read or understood in one particular way. Hall has carefully explored the encoding process, arguing that even as media producers attempt to control the meanings they embed in their messages, that process is always incomplete or partial (Hall 1980). (Consalvo 2005, 8)

Grau (2003, 306–7) notes that "[m]eaning is produced within the systemic structure created by the artists only through the activity of the users involved." For Eco, "*every work of art is potentially 'open,' since it may produce an unlimited range of possible readings*" (Bishop 2004, 62). Brey (1998) observes that

> social structures are malleable, and adapt to technological change in ways that are not always predictable. Newly introduced technology brings along opportunities and threats, to which people respond in ways that may change the surrounding social structure, if not the technology itself, and thereby affect its impacts. … As a result, the political properties of technologies are not fixed but depend on social responses.

Users have profound control over their media. They can use them to their (individual, subversive, etc.) ends, and they say what things mean and what they take them to be: it is always up to people to appropriate media in particular contexts; "[T]he meaning of a [medial] text is always indefinite." (Denzin 1989, 144 qtd. in Winter 1995, 116).

This can be demonstrated in play. "Which of the 'virtual' tendencies become actualized is not directly inscribed in the game's technical properties. They are the 'possibilities opened up by cyberspace technology, so that, ultimately, the choice is ours, the stake in a politico-ideological struggle' (Žižek 1999, 123). (Raessens 2009, 33) Players create meaning, and they assign it to mostly trivial and "blank" game objects and actions (Seeßlen and Rost 1984, 213, my transl.; see Winter 1995, 108). The process in which this occurs is the process of play, and it is very distinct or even completely independent of ordinary life: *Bedeutung im Spiel hat, was Bedeutung für den Spieler erlangt*. Players—not authors or designers—say what things mean in games (Crogan 2011, 174).

Although all media create internal meaning, they appear to differ in the importance references into the external world. For instance, play appears to be more independent of everyday life than film. It is arguably more difficult to understand a movie than a game from a different culture or era, but games can be understood in themselves (Aarseth 2007). This is the case because meanings in play are not or are only to a very limited degree imported from ordinary life. Actions and objects are not relevant in play because they reference play-external actions and objects, but because they become meaningful *inside play*. That is, a football does not stand for anything else; it is not a reference. "The incidence of play is not associated with any particular stage of civilisation or view of the universe. Any thinking person can see at a glance that play is a thing on it[s] own." (Huizinga 1955, 3)

Media are not transparent or immaterial copies of reality; they are entities in themselves. Bagiñski (2005) observes that "[i]t's not possible to make a film without cheating." The medial representation is not a fake because of technical or moral deficits, but in the first place, because *it is a representation*.

> Fictional cinema, as we know it, is based upon lying to a viewer. A perfect example is the construction of a cinematic space. Traditional fiction film transports us into a space: a room, a house, a city. Usually, none of these exist in reality. What exists are the few fragments carefully constructed in a studio. Out of these disjointed fragments, a film synthesizes the illusion of a coherent space. (Manovich 2000, 138)

Saying that media lie is saying that medial representations are different from the original. What spectators see (e.g., in a movie), is not what actually happened, but usually, a very sophisticated illusion. And, as Derida, Jameson, and Baudrillard point out, it means nothing (Winter 1995, 43).

3.2 Active Process

Media reception is not a "passive assimilation of the media product," but an "active and creative process of interpretation and assessment" (see Bisky and Wiedemann 1985, 64ff.), "in which the very meaning of a medial text and its potential enjoyment … are only constructed" (Winter 1995, 115, my transl.). McLuhan's (2002, 367–8) observation that "TV is above all a medium that demands a creatively participant response", is echoed by Manovich (2000, 71–2) who notes that media use is characterized by "psychological processes of filling-in, hypothesis forming, recall, and identification, which are required for us to comprehend any text or image at all"— "[l]ike a postserial musical score which makes a performer into its co-author, 'text' 'asks of the reader a practical collaboration' (see the sixths of the "seven propositions" in Barthes 1977)" (Manovich 2000, 152). Scheuerl (1965, 161, my transl.) notes that "although the spectator [of a theatrical play] does not play [in it], he is (virtually) playing his own game. Even when he does not participate in the performance, he needs to reproduce all images for himself once more on the level of appearance," interpret, combine, think ahead, "to understand anything at all." Although he is not able to grasp the play in its entirety, "he needs, as much as he can, to virtually traverse the area of the play in all dimensions, to inwardly re-enact all movements," if he is to engage properly with the work. Wilson (2003) observes that even "non-interactive media" such as

> a photograph, painting, movie, book, or symphony ... may not be as non-interactive as they appear. Although their form is fixed, the act of engaging them can be highly interactive. ... For example, the reader of a novel or the viewer of a movie is constantly adjusting attention, internal references, identifications, emotional responses, and willingness to engage internal associations that come from personal experience, social/ethnic/gender positions, previous experience with the art form, etc. Some analysts would go so far as to claim there is no successful art or media without this level of engagement interactivity.

This active and participatory orientation towards the world is not limited to media use. All cultural and natural forms immediately, naturally, and intuitively trigger meaning: "Individuals do not passively absorb symbolic forms but creatively and actively *make sense of them*, and thereby produce meaning in the very process of reception" (Thompson 1990, 153 qtd. in Winter 1995, 115).

Media users produce themselves what they experience, including unintentional body movements (e.g., the Carpenter effect). Even in media that invite mental engagement only (e.g., movies) participation appears to be thoroughly embodied:

> Whilst it is true that TV and cinema audiences cannot control the outcome of the scenes they witness, one can suggest that they are not entirely passive throughout the experience. Films can induce a physical response in their audiences (e.g., an adrenalin rush, a release of endorphins, the shedding of tears – either from induced sympathy or from sentimentality). (Mount 2002)

Generally speaking, "[T]he user's body is the first context that the digital designer must take into account" (Bolter and Gromala 2003, 129).

3.3   Decoding Medial Texts

Every medium has its particular code that the reader needs to be able to recognize and to understand. Often, medial codes include considerable abstractions and translations.

A media user knows about her using a medium. This is not a problem or fault, but prerequisite for engagement to occur. Usually, openly announcing or hiding a medium is neither necessary nor possible. Media become transparent and intransparent in use. "As designers, we want the interface to disappear for the user for part of the time, but not completely and not irrevocably. At some subliminal level, the user must be aware of the interface at all times" (Bolter and Gromala 2003, 53). Engagement with media requires and offers both, to be aware of the medium and to get lost in it. This is no contradiction, not a problem to overcome, and not a question to decide (see the section "The Myth of Transparency" in Ibid.; 48ff.). Media users connect and are able to cope with the "tension" (Ibid.:53) between both poles.

In this sense, players are aware that they play, while they play (Huizinga 1955, 23; see Schröck 2005, 50 and Salen and Zimmerman 2004, 526; for a similar observation in theatrical plays and movies see Arnheim n.d., 25). Players play games and manage them formally: "In play as we conceive it the distinction between belief and make-belief breaks down" (Huizinga 1955, 25). Players "step … out of 'real' life into a temporary sphere of activity with a disposition all of its own" (Ibid.:8), but they retain "at the same time … an acute perception of the constraints of the game world and game time and an ability to play strategically within its constraints" (Douglas and Hargadon 2001, 163). This duality also poses a special challenge and thrill (see the chapter, "Playing on the Edge" in this volume): "[I]n the play-mode the deep fascination lies in the oscillation between play and non-play" (Walther 2002). Play is indeed simultaneously real and make-believe. Freud asserts that "[t]he opposite of play is not seriousness, but—[the] reality [of ordinary life]" (Freud 1982, 171f. qtd. in Retter 2003, 62). This is even more surprising in play than in other media, because participants in play change and control its course and outcome; when they act in the play world they do not damage it or destroy its illusion, but on the contrary, build it up and maintain its existence. At the same time, play actions are by all means real, serious, and tangible actions (Sommerseth 2007, 766), often performed with considerable skill, effort, and commitment. A game result cannot be denied.

By its very existence, play creates a distinction between itself and everyday life. This is a peculiar function that can in a similar manner be observed in art: From a Heideggerian perspective, in art "*appearance happens, and it openly announces itself as happening*" (Lewis and Staehler 2010, 104; cf. Heidegger 2001). Art shows the earth "in a way that remains outside of any worldly horizon" (Lewis and Staehler 2010, 104); artworks "withdraw from this world. This is to say that they *appear*, but in a *non*-worldly fashion" (Ibid.). Artworks make other (e.g., past) worlds appear, which stand in contrast or even conflict to the "world in which the artwork actually exists" (Ibid.:105). In the artwork, "appearance *itself* appears" (Ibid.). In this way, it makes people aware of this world without a breakdown of the ready-to-hand that is usually assumed to be a precondition for this experience.

3.4 Media Hint at Themselves

Media are limited, focused, and selective in their representational repertoire. The dream of achieving a perfect medial copy of reality has not been given up only after, for example, the Renaissance, but it was always a Fata Morgana:

> Like other media since the Renaissance—in particular, perspective painting, photography, film, and television—new digital media oscillate between immediacy and hypermediacy, between transparency and opacity. Although each medium promises to reform its predecessors by offering a more immediate or authentic experience, the promise of reform inevitably leads us to become aware of the medium. Thus, immediacy leads to hypermediacy. (Bolter and Grusin 1999, 16 qtd. in Salen and Zimmerman 2004, 452)

Media possess this dual character, otherwise they would not be useful or usable media at all.

The typical Hollywood movie seeks to "kidnap" the audience and to "overwhelm ... [it] by the physical presence of the image" (Sontag 1996 qtd. in Biskind 1999, 404–5). Pauline Kael observes that directors such as George Lucas and Steven Spielberg are "infantilizing the audience, reconstituting the spectator as child, then overwhelming him ... with sound and spectacle, obliterating irony, aesthetic self-consciousness, and critical reflection" (Biskind 1999, 344). But whereas the mainstream of popular media often refrains from using self-references, examples of media indicating their own nature as media abound: for instance, movies by Robert Altman (Biskind 1999, 81–2), Woody Allen's Everyone Says I Love You (1996), and Wayne's World (1992). Several self-references can be found in Harriet Beecher Stowe's *Uncle Tom's Cabin*, first published 1851–1852. Bertold Brecht's *epic theatre* comes to mind.

3.5 The Observer Is Always Participant

Participants can only step back and become observers of their own actions to a limited degree.

> One of the most fundamental aspects of Heidegger's discourse is his emphasis on the state of thrownness as a condition of being-in-the-world. We do at times engage in conscious reflection and systematic thought, but these are secondary to the pre-reflective experience of being thrown in a situation in which we are always already acting. We are always engaged in acting within a situation, without the opportunity to fully disengage ourselves and function as detached observers. (Winograd and Flores 1986, 71)

Salen and Zimmerman suggest that "experience is participation" (Salen and Zimmerman 2004, 314).

Players are part of the game. They are not spectators or external observers, but occupy a limited and privileged position inside the sphere of play; acting in it, changing it, and perceiving it at the same time (Rötzer 2005, 105). For Régis Debray (Robben and Cermak-Sassenrath 2006/7) not only is a player playing a game, but the game is playing the player; a player "has" a game only to the extent that the game "has" the player; they are not standing on opposing sides, or even on the same side, but they are creating each other in a reflexive process.

Werner Heisenberg notes that observations in physics depend on the apparati with which they are made (Ibid.); Niels Bohr asserts that the instruments define what we can measure and observe (Friedman 2013). Reality is created, in an active and shared process. *There is nothing* that could be uncovered or discovered: "I believe that a documentary [film] should not expose reality, it must articulate and structure reality" (Hartmut Bitomsky qtd. in 2012).

This observation applies to all media, not only to interactive ones such as the digital computer or play. The only experience people can have is a "first person experience" (Draper and Norman 1986, 3), even if only they "read or watch a play or movie" (Ibid.; see Laurel's notion of first-personness (Laurel 1986, 76, 77–9, 80 and Norman et al. 1986, 490–1).

3.6 Shared Practice

Players produce meaning inside play, in a shared process over time; actions and objects *become* meaningful as Wittgenstein says of words in a language. In this sense, meaning is an aspect of everyday use and common practice by the people concerned: "Meaning ... does not reside in the [interactive] system itself, but in the ways in which it is used" (Dourish 2001, 183). For Heidegger and Wittgenstein "meaning is embodied in practice, in action in the world" (Ibid.:184). The results are often heterogeneous (Winter 1995, 220), and meanings are constantly adjusted, adapted, and changed on the fly, according to what (different) people do. Meaning cannot be assigned in an abstract fashion, or for other people.

Actions or objects in themselves do not mean anything. Meaning is ascribed by humans (as "intentional actors"), and constructed through (their) actions, in specific settings (Dourish 2001, 184), in a collective fashion (Ibid.:186). This holds true in similar ways for ontology, intentionality, and intersubjectivity. Meaning is not a static or material property but continuously produced in a dynamic and mental process (see the chapter, "Crafting Through Playing," in this volume). "Meaningful play is ... measured by what a player experiences, not by the underlying rules of a game" (Salen and Zimmerman 2004, 226). Learning a game is not to memorize the rules, but to recognize and accept the particular set of "assumptions, conventions, and practices" (Dourish 2001, 186) that players of this game share. By moving from the periphery to the center of this community one recognizes and learns not only to exercise skills, but also to exercise them as a member, and with certain understandings, expectations, significances, and perspectives, that is, meanings (Ibid.; see the chapter, "Little Big Learning: Subversive Play/GBL Rebooted," in this volume).[3]

The meaning of a medial text is constructed in a process of interaction between the text, the reader, and the context. "Popular texts are neither fixed nor closed but always open and contain a potential of meanings. The meaning they assume is not arbitrary, but depends to a considerable degree on the cultural milieu of the readers" (Winter

---

[3]When meaning is essentially an aspect of context (see Suchman 1987, 119), the popular critique of violence in games needs to be reconsidered. Violence is not a category of play. Play is to the same degree violent as, for instance, nature is brutal. Violence is a play-external notion. Game actions have abstract meanings within but not without play. They can only be inadequately judged by everyday rules and ideals.

1995, 222, my transl.). Bennett and Woollacott (1987, 262) point out that "neither text nor context are conceivable as entities separable from one another" and propose to study "the living live of text" in the context of social processes (Winter 1995, 105). The production of meaning appears to be influenced more by the relevance something has for a person in a particular situation than by that person's preference (Ibid.:108).

Play occurs in an abstract and artificial sphere distinct and different from the contexts of ordinary life. The construction of meaning solely depends on the players' decisions and agreement, that is, on their play.

3.7 Coupling

Usually, only relevant things, that is, things that contribute to play are tolerated in it[4]: "It is normally considered a bad design if information is included in the game that does not contribute to one's understanding of what is going on" (Dunningham 2000, 109 qtd. in Salen and Zimmerman 2004, 443, see Laurel 1993, 73ff.). Many fields, pitches, and boards are virtually empty, for this reason. Play is quite limited and focused in this regard, and distinct from ordinary life; it draws solely upon itself to define what things mean.

> The meanings that games produce, while intertwined with larger cultural meanings, still acquire their distinct identity as game meanings because they emerge out of the system of relations made possible by the game itself. This system of relations is not something that naturally occurs out in the world. (Salen and Zimmerman 2004, 572)

Play is heavily self-referential and self-reliant. The meanings of play lie in play itself. "To play a game is to take part in a complex interplay of meaning" (Ibid.:452).

Play actions become meaningful in the process of playing, through an immediate, obvious, and tangible coupling of action and response (see Dourish 2001, 138ff.):

> *Meaningful* play in a game emerges from the relationship between player action and system outcome; it is the process by which a player takes action within the designed system of a game and the system responds to the action. The *meaning* of an action resides in the relationship between action and outcome. (Salen and Zimmerman 2004, 156–7)

Coupling is an extremely effective mechanism to engage the player in the activity of playing (Murray 1997, 146). If play actions do not exhibit such impact and consequence, play might become pointless and meaningless, merely a fooling around: "Arbitrary play, in which actions seem unrelated to each other, is the opposite of meaningful play" (Salen and Zimmerman 2004, 157).

---

[4]Such as the chain saw in *Maniac Mansion* (1987)?

4 Play Is Not a Reference

Play is strongly autopoietic. It is self-referential and only interested in itself. It is conceptually distinct and clearly divided from ordinary life. The meanings, actions, and objects acquired in play are not imported from the outside but created in and through play itself. Play freely uses ordinary actions and objects and pulls them into play. In this process they are stripped of their everyday meanings. Play is not *about* anything; for instance, it does not represent or "picture" (Dourish 2001, 123) ordinary life; it is not about reality. Play is not an abstraction of life or a duplicate or a copy of the everyday world. Although play does not aim to change the world, effects of play often spill out into ordinary life as by-products. As in other media, this spill is avoided, tolerated, or invited by players.[5]

4.1 Conceptual Distinction of Play and Nonplay

For Huizinga (1955) and Scheuerl (1965), play is conceptually clearly divided from everyday life. Among the characteristics they use to describe play, the freedom and the secludedness or limitedness of play appear to be the most relevant for the question at hand.

4.1.1 Freedom of Play

Freedom is an essential characteristic of play: Play is free (Huizinga 1956, 15). Without freedom, play is not imaginable. For Scheuerl, "freedom is among the characteristics of play which constitute it" (Scheuerl 1965, 195, my transl.). Freedom in play can be realized in three ways: the players' free decision *to play*, freedom from ordinary life *in play*, and the players' free decisions *during play*.

Here, the freedom players gain from the everyday world appears to be the most relevant: play is free from ordinary life. This freedom is won through play and in play. Inside the magic circle of play everyday life loses its relevance, significance, and meaning; players become free from it. For the duration of play, the power of ordinary reality is defunct and out of action. Play is not seen as a tool to effectively remove reality, "to free oneself from need or necessity, but as a jubilant expression that one is free already" (Ibid.:74, my transl.).

4.1.2 Secludedness or Limitedness of Play

All forms of play require boundaries (Scheuerl 1965, 95). Fröbel (1937, 16f. qtd. in Scheuerl 1965, 95, my transl.) notes that "play ought to move freely only inside of a

---

[5]Parts of this section have been published in Cermak-Sassenrath (2013).

certain limitation and boundary if it is to truly delight and to satisfy mind and spirit [*Gemüt und Geist*]" (see Buytendijk 1933, 118). Huizinga (1955, 9–10) defines the "secludedness" or "limitedness" as his "third main characteristic of play":

> Play is distinct from "ordinary" life both as to locality and duration. … It is "played out" within certain limits of time and place. It contains its own course and meaning. … More striking even than the limitation as to time is the limitation as to space. All play moves and has its being within a play-ground marked off beforehand either materially or ideally, deliberately or as a matter of course. Just as there is no formal difference between play and ritual, so the "consecrated spot" cannot be formally distinguished from the play-ground. The arena, the card-table, the magic circle, the temple, the stage, the screen, the tennis court, the court of justice, and so on, are all in form and function play-grounds, that is, forbidden spots, isolated, hedged round, hallowed, within which special rules obtain. All are temporary worlds within the ordinary world, dedicated to the performance of an act apart.

Silverstone (1999, 60 qtd. in Kücklich 2006, 12) emphasizes the creation of meaning in play independent of external reality: "Play is a space in which meanings are constructed within a shared and structured place, a place ritually demarcated as being distinct from, and other than the ordinariness of everyday life." The sphere of play ends at the edge of the game board. When *MotoGP* rider Marco Simoncelli states that he will be "arrested" if he collides with Jorge Lorenzo again in a race (Crash Net 2013), he plays with the rider's awareness of the different and disjunct rule systems that regulate ordinary life and play. Clearly, he is not afraid of being arrested by the police during a motorbike race, because he is playing, and nobody from the outside can touch him: play is "reglemented, but not through the moral rules and legal regulations of everyday life" (Krämer 1995, 225–36 qtd. in Pias 2002, 158, my transl.).

Although play is of paramount importance to its players, they are well aware that what they engage in is a game; for them, it is different and distinct from ordinary life. There is a baptism sequence in *Bioshock Infinite* (2013) that the player has to go through in order to proceed with the game. A small number of players expressed that they were not happy with religion being mixed up with a computer game, and one player apparently asked for and received a full refund from the publisher. Summing up the understanding of the relationship between play and ordinary life which is arguably shared by many players, WuShogun212 (2013) comments in a *YouTube* video on this. He calls the complaint "the dumbest thing [he] ever heard in [his] life, it really is the dumbest thing, as a gamer, [he's] ever heard in [his] life, in the gaming industry … [the] dumbest shit [he's] ever heard … [and] ever seen in [his] life." He backs up his position by arguing that everything in games has the same standing towards everyday life, for instance, shooting people and baptisms. He notes (addressing the person who asked for the refund) that "[i]n the game you don't wanna get baptized because you feel [that's] going against your religion, but you're playing a game [in which] you fucking shoot people in the *face* with *guns*." For WuShogun212, games are "not even *real*" and everything in them is "fake," "digital," or "polygons": "[T]he baptism at the beginning of the game … has nothing to do with Christianity at all, yes it's a ritual in Christianity, but it's not a Christian thing in the game, it's not." On the refund the publisher Valve has reportedly offered to the offended player, he comments that "[he doesn't] know why [it was offered]. *[He]* wouldn't have done that." He does not "get why people are so sensitive of something that's not even *real*." In his view, a refund was unnecessary because a game is "something completely different" from ordinary life.

Part of the attraction of play appears to lie in its complete separation from everyday life. An example are social roles or positions in society that lose all relevance in play; it is of no importance if a player is president or student, an "academic, ... an electrician ... [or] a policeman ... " (Röhrbein and Hanke 2006, my transl.), or even if the players feel sympathy for each other, and so on (for a different position see Salen and Zimmerman 2004, 462; cf. 509). The roles of players in a game have "nothing to do with the existing departmental, spatial, economic, or authoritative relationships among players" in the ordinary world (Ibid.:583). If somebody scores in basketball, his team gains points, and he might prove that he is a good player. In play, all players are equal, regardless of what they are outside of it.

Huizinga introduces the notion of the magic circle of play to describe the special, distinct, and delimited sphere of play. The players create this sphere through their play. Conceptually, play constitutes a space untouched by ordinary life, in which rules have absolute authority, and in which players act freely. Players have "the feeling of being 'apart together' in an exceptional situation, of sharing something important, of mutually withdrawing from the rest of the world and rejecting the usual norms" (Huizinga 1955, 12). Playing is not referencing reality (or other media); on the contrary: play needs to distinguish itself. "Inside the circle of the game the laws and customs of ordinary life no longer count" (Ibid.). The magic circle of play

[I]s enclosed and separate from the real world. As a marker of time, the magic circle is like a clock: it simultaneously represents a path with a beginning and end, but one without beginning and end. The magic circle inscribes a space that is repeatable, a space both limited and limitless. In short, a finite space with infinite possibility. (Salen and Zimmerman 2004, 95)

Huizinga's notion of the magic circle describes a place beyond ordinary life. "In the play-state you experience a *protective frame* which stands between you and the 'real' world and its problems, creating an enchanted zone in which, in the end, you are confident that no harm can come" (Apter 1991, 15 qtd. in Salen and Zimmerman 2004, 94). The boundary of play which is established in the heads of the players can coincide with, for example, lines on the floor or special uniforms, or rather, is articulated or realized through them.

Although this frame is psychological, interestingly it often has a perceptible physical representation: the proscenium arch of the theatre, the railings around the park, the boundary line on the cricket pitch, and so on. But such a frame may also be abstract, such as the rules governing the game being played. (Ibid.)

4.2   Play Is Self-Referential

Play is only interested in itself. There are many examples in which, for example, issues, problems, or differences players have outside of play do not translate into play. On the possibility of hiring drummer Matt Sorum into Guns N' Roses, Slash recalls: "The pay was good and there were no rules, except for one: all you had to do was play well" (Slash and Bozza 2008, 309). Hot rod and motorbike customizer Cole Foster states that he does not have any tattoos, no mobile phone, and no *iPod*, and that he did cry when Bambi was shot—but in the end it is only relevant what he builds (Cole Foster in Klimpke and Behlau 2013).

The meanings play creates by and in itself massively outweigh their import from the outside.

> All PvE [player vs. environment] play [in MMOs (massively multiplayer online games)], whether practiced by individuals or groups, remains a meaning-making process that determines values and meanings for game objects. … These values are continuously weighed and refined with reference to the consequences of in-game interactions, yet they can at times – such as when narratives are imposed on game play – also include values and meanings imported from external sources. … In most cases, however, character values and meanings correlate closely with in-game performances. (Myers 2010, 121)

Play does not rely on or wait for ordinary life to be justified or confirmed: "[T]he action [of a contest] begins and ends in itself, and the outcome does not contribute to the necessary life-processes of the group. … Objectively speaking, the result of a game is unimportant and a matter of indifference" (Huizinga 1955, 49). Play is gloriously ignorant of the world like the Long Island community of "West Egg": it is "a world complete in itself, with its own standards and its own great figures, second to nothing because it had no consciousness of being so" (Fitzgerald 2000, 100). Willke (1987, 34, my transl.) echoes von Foerster and describes the autopoietic quality of human—social or personal—systems and claims that they need to be understood as "*closed systems which create and maintain themselves*, that is, they produce their specific dynamics not only as a reaction to input from its surroundings, but primarily through their self-organisation" (Retter 2003, 84). Maturana (1970, 5 qtd. in Myers 1999, 154) terms this a "closed causal circular process." Maturana and Varela (1980) label "this particular formal structure … 'autopoiesis' (or self-forming)" (Myers 1999, 154).

Play is not a copy of (play-external) reality, but creates a reality. It is only itself. "When for example I play at cricket, what am I pretending to do other than the thing which I do?" (Bradley 1906, 468 qtd. in Scheuerl 1965, 84). Play might or might not use actions and objects from everyday life, but meaning is assigned only within play: To enter play "is to move into the magic circle, to move from the domain of everyday life into a special place of meaning. Within this special space the player's experience is guided by a system of representation that has its own rules for 'what things mean'" (Salen and Zimmerman 2004, 366). Meaning is created in play and is not imported from the outside: "No matter how integrated into culture games might be, there will always be some aspect of a game's operation that relies on its own system, rather than that of culture, to create meanings for players" (Ibid.:585). Play rotates around itself, above and beyond the material it plays with; it is performed for its own sake (Scheuerl 1965, 206). This is not unlike art, which is also first and foremost different from other things and is *itself.*

Play also plays with its distinction from ordinary life. It draws actions and objects that players know from specific situations into new contexts. Occasionally, the meaning something has in play is an edgy twist on the meaning something has in ordinary life. Party games (e.g., kissing) and the *Makin' Cake* installation play with the different meanings things have within and without play (see the chapter, "Playing on the Edge," in this volume).

4.3   Import of Objects and Actions into Play

Retter (2003, 99) observes that the play-external world only offers material to play, and that play transforms objects when it moves them into play. The everyday meaning something has might serve as a starting point for people to understand what something means in play, but this meaning is only a weak indication of the meaning something has in play. "The question is … which is to be master—that's all" (Carroll 1993, 205). Play can assign meanings freely and arbitrarily, regardless of what things mean elsewhere. "Game actions refer to actions in the real world, but because they are taking place in a game, they are simultaneously quite separate and distinct from the real world actions they reference" (Salen and Zimmerman 2004, 449). Even more: "When we play a game, we are doing more than just shuffling signs drawn from the domain of the real world; instead, we are shifting to another domain of meaning entirely" (Ibid.:369).

Everyday actions and objects used in play may give it a setting or location, but not its meaning. Naturally, there are quite strong similarities between actions and objects in ordinary life and in play, but this does not indicate play was mimicking reality. The play-external world is not present in play. Players *play for* the meanings in play; they do not *transfer* them from anywhere else. Many games are accurately modeled on situations from everyday life, but this is coincidence: play could easily find other material with which to play. It appears doubtful to conclude that "video games imitate life" (Robinett n.d. qtd. in Salen and Zimmerman 2004, 421).

If "[a] video game is a simulation, a model, a metaphor," it is so only in a very limited sense.[6] Play is not a replacement for reality, and play does not cling to it. Play actions mean something (else) and are purely abstract. Specifically, play is not based on simulation; there are games that also happen to be simulations, such as *SimCity* and flight simulators, but simulation is not a characteristic of play. Play might or might not simulate something just as it might use other materials available. Naturalistic settings or representations, such as animations in *Battlechess*, initially

---

[6]Pias (2013) notes a "partial autonomy" of simulations with regard to, for example, physical laws, such as Awakawa improving weather prediction by disregarding thermodynamic laws in his models.

Fig. 8 Poster advertising the university's inline hockey team (University of Bremen, 2009/10)

offer a certain thrill to players, but lose it the more (e.g., competitive) play takes over. A realistic appearance is conceptually irrelevant for play (e.g., in Brecht's theatre).

Play is a mind game. Whereas "a historical wargame … simulates … the starting conditions of a conflict," and "[t]he way … the conflict plays out … makes the game interesting as a game experience," I would not follow the assumption that "[t]he meaningful play of a historical wargame derives not only from the strategic complexities of military decision making, but also from the fidelity of the game to its historical referent" (Salen and Zimmerman 2004, 442). A story provides only a starting point and background to play: "The historical aspect of these [strategy] games is just the icing on the cake," said Graham Somers, a 22-year-old college student in Vancouver who runs an Age of Empires fan site called HeavenGames. "I have a definite love of history, and certainly sending an army of knights and battering rams into an enemy town has a historical basis, but the main thing is it's a lot of fun. They are games, after all" (Civilization III 2007). The "fidelity to [a] referent" (Salen and Zimmerman 2004, 455) is unimportant for play and strictly optional.

Whether the material that is played with is imported into play, especially manufactured for play, or used exclusively in play, it is only played-with. It is not the content in itself that is interesting to play. Play is not about soccer balls, chess pieces, playing cards, toy trucks, tennis rackets, or princesses. "It's not the things you do … [b]ut it's the way you do the things" (Paragons 1967). The play material (Fig. 8) is, essentially, exchangeable.

The activities of play are often quite ordinary, tedious, or trivial. Although the actions and objects of play might appear similar to actions and objects of everyday life, they lose essential parts of their meaning; play takes away the need, the necessity, and the like. They enter into a new and different space in which they are assigned new meanings independent of the ordinary world, by the players, according to the possibilities they offer to play. Their original meanings might still vaguely hover in the background, but they are not relevant to play. *NBA* player Rodman and Keown (1996) calls the basketball "the fucking thing," and echoes McLuhan (2002, 263) who notes that "[O]stensible program content is a lulling distraction needed to enable the structural form to get through the barriers of conscious attention." Play's content is only a tangible articulation of an ideal space, and is called "material props" by Dunne and Raby (2001, 28), and "texture" by Aarseth.

4.4 Creating Reality

It is not the reflection, the learning, or the gain of insight about reality that drives play, but the *creation of reality*. Play limits itself to importing actions and objects from everyday life, assigning independent meanings to them, and using them interactively. Play is not an abstraction of ordinary life or a duplicate or a copy of the everyday world. It applies what Manovich (2000, 183) observes of computer graphics: "[A]synthetic computer-generated image is not an inferior representation of our reality, but a realistic representation of a different reality." John von Neumann declared at the end of the 1940s the end of the era of representation and the beginning of the era of simulation (Pias 2013). Play is not ordinary life, but real nonetheless. Who wants to say what is real and what is not? Following Luhmann, "[b]oth non-play and play are 'realities,' because they are products of a distinction, a difference that makes a difference" (Walther 2002). In play, players encounter a fully valid reality (cf. Salen and Zimmerman 2004, 449). Play is made up of real actions in unreal worlds. Play is neither a *schein* reality nor an *ersatz* reality; it is not defined or legitimized through references into an external reality, very similar to art which, according to Heidegger, "does not *depict* or *represent* the world. It *creates* a world of its own" (Lewis and Staehler 2010, 103; cf. Heidegger 2001, 44).

In play, players have the real thing. Games are not simulations (see above).[7] Play does not simulate or model the world or other media, but offers possibilities to create reality. It is not only a representation, but has its own dynamic. Play creates reality not because it references everyday life, or because of its (proposed) role as experiment or test environment. Play creates its own fully valid world with meanings independent of external references or purposes (see the chapter, "Crafting Through Playing," in this volume). Playing is creating reality, by play's own standards, rules, and values, wholly unjustified, not legitimized or controlled from anywhere else. Hutchins et al. (1986, 99) describe such an experience in human-computer interaction (HCI):

---

[7]The connections, overlaps, and differences between play and simulation are not discussed here.

The point is that when an interface presents a world of action rather than a language of description, manipulating a representation can have the same effects and the same feel as manipulating the thing being represented. This is the essence of the "first-personness" feeling of direct engagement.

It appears that play can assume a level of significance and relevance for its players which far exceeds that of everyday reality and other media, such as educational exercises. "One of the great problems with the way most schools are set up is that the children quickly sense that most of the stuff they are asked to do is not 'real,' especially as opposed to optional activities like sports and games, art and music" Kay (1995). Play is not real in the sense that is uses situations, actions, or objects from ordinary life, such as bones, mud, or destroyed US tanks in Iraq (see the chapter, "Playing on the Edge," in this volume). On the contrary: play is real because it rises above ordinary life; it lifts things out of everyday reality into play and assigns its own meanings that are realized through the process of play.

4.5 Medial Bleed

Although play does not aim to change the world, the bleed of effects from play into the play-external world is often at least tolerated by players; in many cases this by-product of play actually appears to be quite welcome (e.g., a reputation or fame). Medial overflow into other media or into everyday life is not limited to play, and a well-known phenomenon (see, e.g. Biskind 1999, 7). Also, meanings trickle from ordinary life into play. Apparently, the well-known British train robbers Biggs, Edwards, Goody, Reynolds, Wilson, and Wheater, who took between 2.3 and 2.6 million pounds from a Royal Mail train in August 1963, played *Monopoly* with real money in their farmhouse hideout after the heist (Marks 2007, 85–6). Although play is conceptually divided from ordinary life, every concretely realized game has the potential for exchange with ordinary life. Feelings of victory and defeat easily cross over into everyday life, and players do not dream their games but perform feats that cannot be denied (Seeßlen and Rost 1984, 37). Gambling for money and strip poker are examples of games that players do not only aim to win inside the game world, but that also have real-life components. Another example of games that have profound connections with and into everyday reality is professional sports.

In the practical realization of play, the exchange with everyday reality can hardly be avoided completely; meanings dripple both ways. But ordinary life loses ground to play in the process. If it gains ground, play ends. Play needs to be able to create its own meanings, and to play its own game without ordinary life getting in the way.

Some game theorists will question whether gambling activities can be rightly thought of as games because they have real-world stakes that constantly threaten to destroy the nonserious, playful quality thought to be a core characteristic of genuine game-based activity. Jesper Juul, for instance, will relegate gambling to the no-man's-land between games and nongames in his diagrammatic depiction of concentric circles of games, gamelike activity and nongame forms (Juul 2005, 43). (Crogan 2011, 30); (See the chapter, "Playing on the Edge," in this volume.)

Play reflects culture and transforms it from within: "Cultural elements from outside the circle enter in and have an impact on the game; simultaneously, cultural meanings ripple outward from the game to interact with numerous cultural contexts" Salen and Zimmerman (2004, 572, see 507 and 528). Play not only statically reflects, represents, or depicts culture but also transforms and creates it. This process is one of interplay. Playful elements can be found in many areas of high and low culture. The postmodern self-understanding of culture is substantially oriented towards play, fantasy, and transcendence. On the other side, not all elements of culture can be described as playful (Retter 2003, 113; see the chapter, "Subversive Gamification," in this volume).

Huizinga emphasizes that culture is itself play.[8] Play and culture do not oppose each other but one continues into the other, and they feed off each other. But he laments that the impulses from play for culture diminish in strength and number, so that play appears to be lost for culture.

5 Modes of Participation

The entry into a game is essentially a mental act, a change from one system of meaning into another. "Only when a player has entered into the magic circle of a game rules imbue game actions with meaning and consequence" (Salen and Zimmerman 2004, 537). Players share the process of playing, what Fink terms the "magical production of a play world" (Fink 1957, 35, my transl. qtd. in Retter 2003, 37). This is predominantly an active process, not a reflective one; Heidegger's notion of thrownness applies. Playing is *doing it*, a configurative practice, pure, direct, and immediate action.

Spectators do not play (the same game as players). Their position and their mode of participation are different, as is their experience. Their primary perspective is one of distanced and possibly critical reflection; they relate, consider, and compare. Spectators differ in their readings of medial texts, and these interpretations can lead to struggles about what things mean.

Even more pronounced, there is a potential tension or conflict between players and spectators, because of their different roles and ways of involvement: they understand games differently. The *Makin' Cake* installation plays with these different perspectives and systems of reference. Spectators become players and get involved; players stop playing and begin to reflect on the game.

---

[8]On a pragmatic note, and "[f]or the purposes of game design," (Salen and Zimmerman 2004, 508) "understand 'culture' to refer to what exists outside the magic circle of a game, the environment or context within which a game takes place."

5.1 Action and Reflection in Different Media

Media are biased towards different modes of interaction. Some media primarily allow or require the user to control and shape the action; some media invite reflection (see Winter 1995, 20). One mode of interaction appears to inhibit the other; that is, very little reflection happens inside an ongoing (action) game, and critical distanced reflection almost prohibits a person's own involvement.

Media use can roughly be categorized as either primarily active ("participatory") or reflective ("vicarious") (Newman 2002, 415 in Kücklich 2006, 35, see Murray 1997, 294, note 7). People participate in the action or observe it from the outside. "Watching a film is a passive form of entertainment. When one sits down to watch a film, one expects to be taken on a ride of sorts and be moved in some way by simple observation. Games are different" (Morton 2005). Play is oriented towards external action (see Buchhart 2005): "[O]ur experience of realism in video games is not tied to the perceptive process, understood as the passive reception of visual stimuli, but to the enactive process, to movement and bodily sensation" (Sommerseth 2007, 766).

In play, the question is not, "*What is it?*," but, "*What can it do?*" (Salen and Zimmerman 2004, 87, see Ebert 1997, 8). It appears possible to say that people like to read and watch because they like to reflect; they like to play because they like to act. "Markku Eskelinen ... points out, drawing on Espen Aarseth's well-known typology of cybertexts, that playing a game is predominantly a configurative practice, not an interpretative one like film or literature" (Klevjer 2002 qtd. in Frasca 2003, 4).

Lash (1990, 175ff. in Winter 1995, 56) observes a difference between discourse (theory) and figure (art), and uses Susan Sontag's *Against Interpretation* (1966) description (Poague and Parsons 2000, xlii) of a "new sensibility" to propose a corresponding differentiation between two forms of cultural sensibilities: the modern discursive and the postmodern figural sensibility. In the discursive sensibility, words dominate images; the formal qualities of cultural objects are much valued; culture is seen rationally; the meanings of cultural texts are given special attention; the sensibility emphasizes Freud's ego over his id; and distance to the cultural object is a precondition for its reception. The figural sensibility is more visually oriented than literarily; formal qualities lose importance, and everyday materials are used; it questions rational and/or "didactic" ideas about culture; it is more interested in what a text does, than what it means; Freud's primary process is observed in culture; the reader's immersion into the experience of the text is preferred (Winter 1995, 56).

5.2 Pure Action

Play is active, immediate, and direct (see Keller 1998, 196–7): "Playing a game means making choices and taking actions" (Salen and Zimmerman 2004, 33). All games are about what the player does, not what he thinks, reflects, or imagines; even in (story-based) adventure games this is unquestioned (Pias 2002, 104). "Games

... are primarily about the experience of the moment, and challenging of the self. ... Most games are about remaining focused in the moment, and acting skill-fully" (Glassner 2001, 58). Raessens (2009, 32) notes that (even in serious games such as *Food Force* (2005)) playing is not a reflective experience, at least not during a player's own play: "[T]here seems to be little room for a critical, reflective atti-tude towards the game's ideology while playing these games." Clearly, "[m]echanics trump meaning" (Aarseth 2007). Marc Cohn (1991) sings of a similar experience:

> I know there's gonna be a lesson somewhere
> I'm gonna think a lot about it later
> But right now I'm miles away
> Miles away ...
> I'm a million miles away
> Where I don't have to think at all.[9]

Glassner (2001, 58–9) proposes that the difference in a person's involvement in stories and games lies in the different kinds of mental engagement: readers "consider the situation" and "weigh the consequences," players execute "immediate ... action" and "there's no time for deliberation." Similarly, Huhtamo (2005, 67, my transl.) observes that "[i]n computer games the player is caught in activity, and cannot reflect [*kann nicht zum Nachdenken kommen*]." Rötzer (2005, 106, my transl.) describes "the computer game" as "the starting point of a new aesthetic that is based not on a distant, passive recipient but on an involved, acting player [*(Mit)Spieler*]."

Playing appears to be a trip and can reach intense levels of immediacy and impor-tance:

> When a tennis player is "on his game," he's not thinking about how, when, or even where to hit the ball. He's not trying to hit the ball, and after the shot he doesn't think about how badly or how well he made contact. The ball seems to get hit through an automatic process which doesn't require thought. (Gallwey 1974, 31ff. qtd. in Pias 2002, 85)

Racing driver von Brauchitsch (1943, 57) describes how in the heat of the moment, when one car attempts to overtake another, the drivers are completely igno-rant of what the spectators on the stands call their "lives." Everything except the race has vanished and has become unimportant and even nonexistent. All the competitors think about and aim to do is to fight and to win, fueled by the sensations of immense power and the thrill of speed.[10] In art, "[c]olor shines and wants only to shine" (Hei-degger 2001, 45 qtd. in Lewis and Staehler 2010, 236, footnote 20). Play is pure action, and players have no second thoughts: "He climbed, as Ochun suggested, as

---

[9]See the chapter, "Questions over Answers: Reflective Game Design and Playing the Subject," in this volume.

[10] In den Momenten des erbittertsten Kampfes aber, wenn hinter oder fast schon neben [den Rennfahrern] der Wagen des Gegners sich heranschiebt, denkt keiner mehr daran, was die anderen, die auf den Tribünen sitzen und zuschauen, ihr "Leben" nennen. ... Das alles ist verschwunden, gleichgültig und wesenlos geworden. Der Wille zu kämpfen und zu siegen, getragen und gesteigert von dem Hochgefühl eines unheimlichen Kraftbewußtseins und dem gefährlichen Rausch irrsinniger Schnelligkeiten, ist einzig und allein restlos beherrschend.

though he were delighted to do so, with no more purpose in mind than proving that he could" (Gibson 2008, 337).

5.3   Relationship Between Players and Spectators

Players decide freely about their actions within play (see Keller 1998, 155), and their decisions cannot be questioned from without. Of course, play is a social situation, but players can free themselves from it. They do not have to answer to people outside play, and are not bound by anybody's interests, ideas, expectations, rules, or customs, be it spectators or coaches. Athletes, for instance, despite being financially supported by the state, remain free to play their own game (Gebauer 2012).

Play can mean everything for its players, and nothing for ordinary life. It is "something tedious and self-importantly arcane. … Something that [does not] matter, [is] of no great importance, on which nothing [depends]" (Gibson 2011, 379). Seen rationally, the result of play is thin air; it only becomes interesting because participants and spectators decide so (Huizinga 1956, 49). Spectators are optional for play (Ibid.) because they have no part in it, and do not even exist in it. Players play; they do not present, perform, show, or entertain. Their job is not the creation of an exciting enjoyable spectacle for an audience. They play purely for themselves.

Who is player, and who is spectator? Whether somebody is playing can only be decided from the perspective of play. A referee does not play, at least not the same game as the players around her. Football fans play their own game when they try to sing louder than the fans of the other team, and so on, but they do not participate in the action on the pitch, despite all propaganda about them being the twelfth member of the team.

Actors and spectators maintain a relationship: an actor plays for an audience, and "the audience creates the magic spotlight in which the actors move" (Murray 1997, 115). Mann (2000, 36) calls this a "*wechselseitiges Sich-Genüge-Tun, eine hochzeitliche Begegnung seiner und ihrer Begierden.*" An actor plays roles, and if his performance is weak it is justified to a certain degree to say that he is a bad player. But a player who loses a game cannot be blamed by spectators; she is not playing for them. It indicates a misunderstanding and is pointless if, for instance, soccer players apologize to the fans for a miserable performance.

5.4   Passing the Boundary

Play happens in a special sphere of meaning that cannot be touched by ordinary life. To enter a game is to know, to accept, and to share its perspective:

> The frame of a game communicates that those contained within it are "playing" and that the space of play is separate in some way from that of the real world. ... Players acting within the frame of the game do so according to rules and the contexts that determine the meaning of those actions. (Salen and Zimmerman 2004, 370–1)

For Silverstone (1999, 60 qtd. in Kücklich 2006, 12),

> Play is part of everyday life, just as it is separate from it. To step into a space and a time to play is to move across a threshold, to leave something behind – one kind of order – and to grasp a different reality and a rationality defined by its own rules and terms of trade and action.

To enter play and to be admitted into it is "to move into the magic circle, to move from the domain of everyday life into a special place of meaning" (Salen and Zimmerman 2004, 366).

> To play a game is to submit your behavior to the rules of the game, to enter into the time and space that the game demarcates, to traffic in the special meanings that the game offers up. To play a game is to participate in the discourse of the game with the other players. (Ibid., 256)

How do players become players? Walking onto a field does not make somebody a player. The sphere of play might or might not coincide with a particular physical space. Although the play world lies in the midst of everyday life and everyday actions and objects are used within it, it is entirely different, special, and impenetrable. The act of entering into it is marked by a decision to leave behind ordinary life and enter play. This transition is a mental step, that is often reflected by, for example, moving into a certain spatial area or wearing a certain type or article of clothing. This change of appearance signals to the other players and to people not playing a change of perspective.

The worlds of play and everyday reality are so distant and different that exchanges across the border usually only happen at specially designated times or in certain situations through a particular ritual. The transition is such an important step that it is often highly regulated. If an exchange needs to be done on the spot, for instance, when a soccer player is injured, it is a major disruption and follows a meticulous protocol. But even in a kids' soccer game in the afternoon, there is a certain procedure for a new kid of proving him capability, offering herself, waiting some time, and finally her being tentatively admitted into play when somebody is tired or has to leave early.

The existence of the magic circle marks the beginning and the duration of play.

> Beginning a game means entering into the magic circle. Players cross over this boundary to adopt the artificial behaviors and rituals of a game. During the game, the magic circle persists until the game concludes. Then the magic circle dissolves and players return to the ordinary world. (Ibid.:333)

Outside of the circle no play occurs. Players who leave it stop playing and enter everyday reality.

The world of play has a limited existence; it might exist only for a short time, such as for 90 minutes or one night. When the referee blows his whistle or the lights are switched off it stops. The players leave the pitch or the stage, exchange their uniforms or costumes for their everyday attire, and find themselves back in their ordinary lives.

What a disenchantment, stepping out of play onto the street. How boring, unimportant, and tedious everyday life feels. Players no longer make brave and essential decisions, perform skillfully daring actions, and fight for honor and glory; they quit the magic world of play of which ordinary life knows nothing.

5.5   Play Is Players

Play needs players, and it is only players who "make decisions that move the game forward" (Ibid.:164). Players bring play into existence; it does not exist without them (see Lamnek 1989, 24f. in Keller 1998, 101; Jordan 2005, 101). The formal rules, the physical space, the tangible toys only help players to bring play about. They are also expressions of play, but they themselves are of course not play: "The board, the pieces, and even the rules of Chess can't alone constitute meaningful play" (Salen and Zimmerman 2004, 33). If it were not for the "shared efforts of the players" the actions and objects would not be imbued with special meaning and importance, and elevated above ordinary reality into "a game's fragile magic circle [which] takes shape and is sustained over the course of play. ... The players cooperatively form the space of the game, in order to create a competition for their own amusement" (Ibid.:256; see Scheuerl 1965, 112f.). Play is an experience. A game is only play for its players. If somebody wants to experience play she herself has to play. Play can only be anticipated in advance or enjoyed through identification to a limited degree.

5.6   Conflict

Action and reflection are not particular activities but cognitive states or modes of participation. A change between an active and a reflective role can also be described as a change from one body or set of possibilities for action into another. Such a change necessitates a re-evaluation in regard to a participant's own contextual stance and position. A change in perspective is reflected or triggered by a change in the mode of involvement.

Games clearly emphasize action over reflection.

> Developers of interactive multimedia need to keep in mind the criticality of deep psychological interactivity of successful art and media. The structural incorporation of concrete choice making does not guarantee deep engagement. Indeed, some analysts suggest that the choice making itself can distract from this deep engagement by disrupting the possibilities of these internal processes of feeling and musing. (Wilson 2003)

Games are decided by rapid rational decisions, plans, and actions; winning is not connected to moral or emotional considerations of the game content, nor to narrative identification with characters or situations (Fritz 1999, 93). In short, "[r]eflection kills" (Timothy Druckrey, pers. comm., Nov 19, 2008; see Grau 2003, 201ff.).

Changes between modes of involvement happen between different media or within a medium. Manovich (2000, 189) discusses cut scenes and assumes these changes are also "typical of ... modern computer use in general. ... The oscillation between illusionary segments and interactive segments forces the user to switch between different mental sets—different kinds of cognitive activity" (Ibid.). In the *Cake* installation, participants change from being spectators to becoming players and vice versa.

Changes do not lead to the cognitive perspectives harmoniously complementing each other, on the contrary: tensions mount between action and reflection, between different people in different roles, and within the same person in different roles:

> Often, the two goals of information access [i.e., action] and psychological engagement [i.e., reflection] compete within the same new media object. Along with surface versus depth [see Ibid.:189], the opposition between information and "immersion" can be thought of as particular expression of the more general opposition characteristic of new media: between action and representation. And just as it is the case with surface and depth opposition ... the results of this competition are often awkward and uneasy. (Ibid.:192)

Action and reflection appear as conceptually different and divergent perspectives on media. Both are powerful approaches and positions on their own, but they refuse to converge or mix gradually in an "area of immersive enchantment" (Murray 1997, 267). Media lean towards one or the other and go all the way. It applies what Adams (1999) observes about interactivity and storytelling: "[T]hey exist in an inverse relationship to one another. The more you have of one, the less you're going to have of the other."

6   Conclusion

Media are open and undefined; their users appropriate them in context. Meaning is not embedded or hidden in medial texts, but is creatively produced and (dis-)agreed upon, depending on the relevance things have for their readers in a dynamic, collaborative, ongoing, and also stable process. The process of the construction of meaning is not abstract, distant, or based on observation, but it is based on actively becoming and being a participant in it, on everyday use and common practice that reflect and constitute meaning in a natural and intuitive way. Actions and objects in games are very basic, and thus offer a vast potential of meaning. It is only through the shared efforts of the players that play acquires and maintains meaning. Play is not dependent on or even interested in the subject matter with which it plays. Although there is an interplay, and play material is important to play, it is only played-with. It does not mean anything, that is, beyond play. Likewise, players' actions do not mean anything. In this respect, play acts as a fun-house mirror into Wonderland, reflecting ordinary life but giving it its own twist, path, and, finally, meaning, free and independent of the everyday world. I propose the notion of the opacity of play to describe this phenomenon. Players accept all kinds of play actions, because they are blank (Seeßlen and Rost 1984, 213) and mean nothing. Every child who plays, "Knows that it plays" (Huizinga 1955, 18) and is aware that it is "only pretending" (Ibid.:22; cf. Salen and Zimmerman 2004, 526; Schröck 2005, 50).

Play is conceptually clearly divided from ordinary life. It is strongly autopoietic; the meanings actions and objects acquire within play are not imported from the outside but created in and through play itself. Players assign meanings to actions and objects that only depend on the meanings they have or gain in play. Other media may reference ordinary life to a stronger degree, or rather, at all. Play is opaque with regard to meaning. It is not an abstraction of ordinary life or a duplicate or a copy of the everyday world. Play is not *about* anything, but play creates. Play does not aim to change the world, however, the effects of play cross over into everyday life as a by-product.

The entry into play is a change from one system of meaning into another. Players share the process of playing and the production of the play world: they collaboratively create the magic circle. This is primarily an active process, a configurative practice, pure, direct, and immediate action. The position of spectators is different. Their primary perspective is one of distanced and possibly critical reflection; they relate, consider, and compare. These different kinds of involvement, understanding, and experience offer a potential for tension or conflict between players and spectators. Players "are different and do things differently" (Huizinga 1955, 12), and most naturally step out of the systems of meaning that surround them in the everyday world, without being social freaks or media experts. Satre is concerned with "the contingent features of a situation and a certain entity, which lend it its unique character, a character which cannot be predicated without actually experiencing that thing from the 'inside'" (Lewis and Staehler 2010, 121). Players understand play, spectators do not, because they have different perspectives on what is happening, and different ways to participate. "It's hard to explain this one, but if you were one of us and did it, then you would understand" (Williams 1988, 104 qtd. in Winter 1995, 125).

This situation is the potential the Makin' Cake installation plays off: when the meaning of a medial text is always open (Winter 1995, 222) and is to be determined in an heterogeneous manner by the people involved, and cannot be predicted, there is potential for conflict. The installation goes a step further to create a confrontational situation in which the restraints, customs, laws, rules, and so on of ordinary life are juxtaposed with the freedom of play. Players and spectators have to face the conflicts that appear between them; but there are also conflicts to face for the same person having been a spectator before and being a player now. *Makin' Cake* emphasizes this change of perspective, this step into and out of play; it confronts people with their own joyful, unreflective, direct, and immediate experience, and provokes people in their role as spectators.

References

Aarseth, E. J. (2007). Semiotics vs. mechanics, or fiction/simulation/reality? In *Two game-ontological models*. Lecture, Informatik-Kolloquium, Univ. Bremen, Dec 5, 2007. My notes.

Adams, E. W. (1999). Three problems for interactive storytellers. *Gamasutra*, Dec 29, 1999. Retrieved Jun 4, 2004, from www.gamasutra.com/features/designers_notebook/19991229.htm.

Anonymous. (1954). Fruit pavlova wins £1/1/. *The Courier-Mail, Brisbane*, p. 11, Oct 13, 1954. Retrieved May 2, 2013, from http://trove.nla.gov.au/ndp/del/article/50624617.

Anonymous. (2009). Storm over call of duty game that allows players to massacre civilians. *Mail Online*, 11 Nov 2009. Retrieved May 2, 2013, from www.dailymail.co.uk/news/article-1226588/Call-Duty-Political-storm-brutal-video-game-allows-killing-civilians-airport-massacre.html.

Apter, M. J. (1991). A structural-phenomenology of play. In J. H. Kerr & M. J. Apter (Eds.), *Adult play: A reversal theory approach*. Amsterdam: Swets and Zeitlinger.

Arnheim, R. (n.d.). *Film as art* (18th edn). Berkeley: University of California Press.

Bagiński, T. (2005). On his animated movie *Fallen Art*. Ars Electronica 2005, September 2005. My notes.

Barthes, R. (1977). From work to text. In *Image-music-text*. New York: Hill and Wang. English translation Stephen Heath.

Bennett, T., & Woollacott, J. (1987). *Bond and beyond: The political career of a popular hero*. London: Macmillan.

Bishop, C. (2004). Antagonism and relational aesthetics. *October, 110*(60), 51–79. Retrieved Apr 8, 2011, from www.jstor.org/stable/3397557.

Biskind, P. (1999). *Easy riders, raging bulls. How the sex 'n' drugs 'n' rock 'n' roll generation saved hollywood* (10th ed.). London: Bloomsbury, paperback.

Bisky, L., & Wiedemann, D. (1985). *Der Spielfilm: Rezeption und Wirkung*. Berlin: Henschel.

Bolter, J. D., & Gromala, D. (2003) *Windows and mirrors: Interaction design, digital art, and the myth of transparency*. Cambridge: MIT Press.

Bolter, J. D., & Grusin, R. (1999). *Remediation: Understanding new media*. Boston: MIT Press.

Bradley, F. H. (1906, October). On floating ideas and the imaginary. *Mind, XV*(60), 445–72.

Brey, P. (1998). The politics of computer systems and the ethics of design. In J. van den Hoven (Ed.), *Computer ethics: Philosophical enquiry* (pp. 64–75). Preprint version: Rotterdam University Press.

Buchhart, D. (2005 Jun–Aug). Über die Dialektik von Spielregeln und offenem Handlungsfeld. *Kunstforum International, 176*, 38–55.

Buytendijk, F. J. J. (1933). *Wesen und Sinn des Spiels: Das Spielen des Menschen und der Tiere als Erscheinungsform der Lebenstriebe*. Berlin: Kurt Wolff.

Carroll, L. (1993). *Alice's adventures in wonderland and through the looking-glass*. Ware.

Cermak-Sassenrath, D. (2013). Makin' cake and the meaning in games. In L.-L. Chen, T. Djaja-diningrat, L. Feijs, S. Fraser, J. Hu, S. Kyffin & D. Steffen (Eds.), *Design and semantics of form and movement (DeSForM)* (pp. 199–203).

Civilization III. (2007). Retrieved January 18, 2007, from www.civ3.com/de/news.cfm.

Cohn, M. (1991). Miles away. LP *Marc Cohn*, lyrics from www.asklyrics.com/display/marc-cohn/miles-away-lyrics.htm (Feb 13, 2014), Written by Paul Taylor.

Consalvo, M. (2005). Rule sets, cheating, and magic circles: Studying games and ethics. *International Review of Information Ethics (IRIE), 4*, 7–12.

Crash Net. (2013). Post-qualifying press conference at Estoril, May 1, 2011. Retrieved May 1, 2013, from www.crash.net/motogp/news/168821/1/verbal_clash_between_lorenzo_and_simoncelli.html.

Crogan, P. (2011). *Gameplay mode: War, simulation, and technoculture*. Minneapolis: University of Minnesota Press.

D1scR3adErr0r. (2013). Does swearing make a better game? *4Player Network*, Apr 28, 2013. Retrieved May 14, 2013, from http://4playernetwork.com/cblogs/D1scR3adErr0r/2013/04/does-swearing-make-a-better-game.

Denzin, N. K. (1989). *Interpretive Interactionism.* London/Newbury Park/New Delhi: Sage.

Douglas, J. Y., & Hargadon, A. (2001). The pleasures of immersion and engagement: Schemas, scripts and the fifth business. *Digital Creativity, 12*(3), 153–66.

Dourish, P. (2001). *Where the action is: The foundations of embodied interaction.* Cambridge: MIT Press.

Draper, S. W., & Norman, D. A. (1986). Introduction. In D. A. Norman & S. W. Draper (Eds.), *User centered system design: New perpectives on human-computer interaction* (pp. 1–5). Hillsdale: Lawrence Erlbaum.

Dunne, A., & Raby, F. (2001). *Design noir: The secret life of electronic objects.* Basel/Boston/Berlin: Birkhäuser.

Dunningham, J. F. (2000). *Wargames handbook: How to play and design commercial and professional wargames* (3rd ed.). San Jose: Writers Club Press.

Ebert, H. (1997). Zu den Bildern von Eva Koethen. In E. Koethen (Ed.), *Bild-Räume 1993–1996* (pp. 8–12). Berlin: Brains.

Ekins, C. (2009). The 1940's experiment—weight loss blog—losing 100 lb on wartime rations. Blog, Oct 14, 2009. Retrieved May 2, 2013, from https://1940sexperiment.wordpress.com.

Fink, E. (1957). *Oase des Glücks. Gedanken zu einer Ontologie des Spiels.* Freiburg: Karl Alber.

Fiske, J. (1988). Critical response: Meaningful moments. *Critical Studies in Mass Communication, 5*(3), 246–51.

Fitzgerald, F. S. (2000). *The Great Gatsby [1926].* London: Penguin.

Frasca, G. (2003). Ludologists love stories, too: Notes from a debate that never took place. In *Digital Games Research Conference 2003 Proceedings* (www.gamesconference.org) *and DiGRA Digital Library* (www.digra.org).

Freud, S. (1982). *Studienausgabe* (Vol. 10). Frankfurt: Fischer.

Friedman, K. (2013). Keynote speech. Creativity and Cognition 2013, Jun 18, 2013. My notes.

Fritz, J. (1999). *Herrscher in virtuellen Welten—die Ausübung von Macht. Spektrum der Wissenschaft, Dossier Software* (pp. 91–4).

Fröbel, F. (1937). *Theorie des Spiels* (Vol. I). Langensalza: Julius Beltz. Introduction Erika Hoffmann.

Gallwey, W. T. (1974). *The inner game of tennis.* New York: Random House.

Gebauer, G. (2012). Deutschlandfunk, Informationen am Mittag, Jul 25, 2012. My notes.

Gibson, W. (2008). *Spook Country.* London: Penguin.

Gibson, W. (2011). *Zero History.* London: Penguin.

Glassner, A. (2001). Interactive storytelling: People, stories, and games. In O. Balet, G. Subsol & P. Torguet (Eds.), *Virtual Storytelling Using Virtual Reality Technologies for Storytelling: Proceedings International Conference ICVS 2001, Avignon, France, 27–28 Sep 2001* (Vol. 2197, pp. 51–60)., Lecture Notes in Computer Science Berlin: Springer.

Grau, O. (2003). *Virtual art: From illusion to immersion.* Cambridge: MIT Press, English translation Gloria Custance.

Guns n' Roses (1988). Used to love her. LP *G N' R Lies.*

Hall, S. (1980). Encoding/decoding. In S. Hall, D. Hobson, A. Love & P. Willis (Eds.), *Culture, media, language: Working papers in cultural studies, 1972-79* (pp. 128–38). London: Hutchinson.

Hall, S. (1993). Encoding, decoding. In S. During (Ed.), *The Cultural Studies Reader* (pp. 90–103). London: Routledge.

Heidegger, M. (2001). The origin of the work of art. (1936). *Poetry, language, thought* (pp. 15–88). New York: Harper and Row. English translation Albert Hofstadter.

Huizinga, J. (1955). *Homo Ludens: A study of the play-element in culture.* Boston: Beacon Press.

Huizinga, J. (1956). *Homo Ludens. Vom Ursprung der Kultur im Spiel.* Reinbek bei Hamburg: Rowohlt. German translation Hans Nachrod.

Huhtamo, E. (2005). Zurücksprechen. Kunstforum. *International, 178,* 63–9.

Hutchins, E. L., Hollan, J. D., & Norman, D. A. (1986). Direct manipulation interfaces. In D. A. Norman & S. W. Draper (Eds.), *User centered system design, new perpectives on human-computer interaction* (pp. 87–124). Hillsdale: Lawrence Erlbaum.

Irrational Games Forums. (2013). The baptism controversy. http://forums.2kgames.com/showthread.php?234121-The-baptism-controversy.

Jordan, J. (2005). Verführung. Kunstforum. *International*, *178*, 101–5.

Juul, J. (2005). *Half-real: Video games between real rules and fictional worlds.* Cambridge: MIT Press.

Kay, A. (1995). Powerful ideas need love too! Written remarks to Joint Hearing on Educational Technology in the 21st Century, Science Committee and the Economic and Educational Opportunities Committee, US House of Representatives, Washington D.C., Oct 12, 1995. Retrieved September 10, 2004, from http://minnow.cc.gatech.edu/learn/12.

Klimpke, K., & Behlau, D. (2013). *Show me your garage. Custombike*, *3*, 64–9.

Keller, P. E. (1998). *Arbeiten und Spielen am Arbeitsplatz: Eine Untersuchung am Beispiel von Software-Entwicklung.* Frankfurt a. M.: Campus, 1998. Also: Bremen, Univ., Diss., 1997.

Klevjer, R. (2002). In defense of cutscenes. In F. Mäyrä (Ed.), *CGDC conference proceedings.* Tampere: Tampere University Press.

Krämer, S. (1995). Spielerische Interaktion. Überlegungen zu unserem Umgang mit Instrumenten. In F. Rötzer (Ed.), *Schöne neue Welten? Auf dem Weg zu einer neuen Spielkultur* (pp. 225–237). Munich: Boer.

Kücklich, J. R. (2004). Play and playability as key concepts in new media studies. STeM Centre, Dublin City University. Retrieved April 19, 2006, from www.playability.de/Play.pdf.

Kunstraum der Leuphana Universität Lüneburg. Email, invitation to the screening of Hartmut Bitomsky's *Reichsautobahn*, (FRG 1985) on Jul 6, 2012, Jun 28, 2012.

Lamnek, S. (1989). *Qualitative Sozialforschung* (Vol. 1) (Methodologie). Munich: Psychologie Verlags Union.

Lash, S. (1990). *Sociology of Postmodernism.* London/New York: Routledge.

Laura, V. (2013) Why unnecessary swearing in video games annoys me. *Leviathyn. The Gamer's Chronicle*, Apr 8, 2013. Retrieved May 14, 2013, from http://www.leviathyn.com/games/opinion/2013/04/08/why-unnecessary-swearing-in-video-games-annoys-me.

Laurel, B. K. (1986). Interface as mimesis. In D. A. Norman & S. W. Draper (Eds.), *User centered system design: New perpectives on human-computer interaction* (pp. 67–85). Hillsdale: Lawrence Erlbaum.

Laurel, B. K. (1993). *Computers as theatre.* Reading: Addison-Wesley.

Lewis, M., & Staehler, T. (2010). *Phenomenology. An Introduction.* London: Continuum.

Mann, T. (2000). *Bekenntnisse des Hochstaplers Felix Krull. Der Memoiren erster Teil* (43rd Ed.). Frankfurt a. M.: Fischer.

Manovich, L. (2000). The Language of New Media. Retrieved Augest 9, 2002, from www.manovich.net/LNM/Manovich.pdf.

Marks, H. (2007). *Señor nice: Straight life from wales to South America.* London: Vintage Books.

Maturana, H. R. (1970). The neurophysiology of cognition. In P. L. Garvin (Ed.), *Cognition: A multiple view* (pp. 3–24). New York: Spartan.

Maturana, H. R., & Varela, F. J. (1980). *Autopoiesis and cognition: The realization of the living.* Dordrecht: Reidel.

McLuhan, M. (2002). *Understanding media: The extensions of man.* London: Routledge.

McNeilly, J. (2008). 10 most swearing-est games. *GamesRadar*, Apr 3, 2008. Retrieved May 14, 2013, from www.gamesradar.com/10-most-swearing-est-games.

Morton, S. B. (2005). Enhancing the impact of music in drama-oriented games. *Gamasutra*, Jan 24, 2005. Retrieved January 27, 2006, from www.gamasutra.com/features/20050124/morton_01.shtml.

Mount, P. (2002). Gameplay: The elements of interaction. *Gamasutra*, Apr 3, 2002. Retrieved April 25, 2003, from www.gamasutra.com/education/theses/20020403/mount_01.htm.

Murray, J. H. (1997). *Hamlet on the Holodeck: The future of narrative in cyberspace*. Cambridge: MIT Press.

Myers, D. (1999). Simulation as play: A semiotic analysis. *Simulation and Gaming, 30*(2), 147–62.

Myers, D. M. (2010). *Play redux: The form of computer games*. Ann Arbor: University of Michigan Press.

Newman, J. (2002). In search of the videogame player. The lives of Mario. *New Media and Society, 4*(3), 2002.

Norman, D. A., Draper, S. W., & Bannon, L. J. (1986). Glossary. In D. A. Norman & S. W. Draper (Eds.), *User centered system design: New perpectives on human-computer interaction* (pp. 487–497). Hillsdale: Lawrence Erlbaum.

Paragons. (1967). The tide is high. Written by John Holt.

PenVampyre@aol.com. (2011). Homemade vintage cake recipes, Dec 24, 2011. Retrieved October 23, 2012, from www.twisted-candy.com/1950s-cake-recipes.html.

Pias, C. (2000). *Computer Spiel Welten*. Munich: Sequenzia, 2002. Also: Weimar, Univ., Diss., 2000. Retrieved October 28, 2004, from ftp://ftp.uni-weimar.de/pub/publications/diss/Pias/pias. pdf.

Pias, C. (2013). On simulation. Talk, ITU, Copenhagen, Dec 13, 2013. My notes.

Poague, L. A., & Parsons, K. A. (Eds.). (2000). *Susan Sontag: An annotated bibliography 1948–1992*. New York: Routledge.

Polygon. Valve refunds gamer uncomfortable with forced baptism in BioShock Infinite. Retrieved April 16, 2013, from www.polygon.com/2013/4/16/4231064/valve-refunds-baptism-bioshock-infinite.

Raessens, J. (2009). Serious games from an apparatus perspective. In M. van den Boomen, S. Lammes, A.-S. Lehmann, J. Raessens & M. T. Schäfer (Eds.), *Digital material-Tracing new media in everyday life and technology* (pp. 21–34). Amsterdam: Amsterdam University Press.

Retter, H. (2003). *Einführung in die Pädagogik des Spiels*. Braunschweig: Institut für Allgemeine Pädagogik und Technische Bildung der Technischen Universität Braunschweig, Abteilung Historisch-Systematische Pädagogik, 1998, new ed. 2003.

Robben, B., & Cermak-Sassenrath, D. (2006/7). Theorie digitaler Medien. Course, WS2006/7, Medieninformatik, Hochsch, Bremen. My notes.

Robinett, W. (n.d.). Inventing the adventure game. Unpublished.

Rodman, D., & Keown, T. (1996). *Bad as I wanna be*. New York: Delacorte Press.

Röhrbein, K., & Hanke, M. (2006). Die wollen nur spielen. *Waldeckische Landeszeitung*, Aug 5, 2006.

Rötzer, F. (2005 Jun–Aug). Die Begegnung von Computerspiel und Wirklichkeit. *Kunstforum International, 176*, 102–15.

Salen, K., & Zimmerman, E. (2004). *Rules of play: Game design fundamentals*. Cambridge: MIT Press.

Scheuerl, H. (1965). *Das Spiel. Untersuchungen über sein Wesen, seine pädagogischen Möglichkeiten und Grenzen* (4th/5th ed.). Weinheim: Julius Beltz.

Schröck, P. (2005). Wir spielen immer. Kunstforum. *International, 178*, 48–57.

Seeßlen, G., & Rost, C. (1984). *Pac-Man & Co. Die Welt der Computerspiele*. Reinbek bei Hamburg: Rowohlt.

Silverstone, R. (1999). *'Play': Why Study the Media?*. London: Sage.

Slash, & Bozza, A. (2008). *Slash* (6th ed.). London: Harper Collins.

Sommerseth, H. (2007). 'Gamic realism': Player, perception and action in video game play. In *Situated Play, Proceedings of DiGRA 2007 Conference* (pp. 765–768).

Sontag, S. (1996). Storytelling metaphors for human-computer interaction in Mixed Reality. *New York Times*, Feb 25, 1996. Retrieved January 15, 2014, from www.nytimes.com/books/00/03/12/specials/sontag-cinema.html.

Suchman, L. A. (1987). *Plans and situated actions: The problem of human-machine communication*. Cambridge: Cambridge University Press.

Thompson, J. B. (1990). *Ideology and modern culture: Critical social theory in the era of mass communication*. Cambridge: Polity Press.

Volta, G. (2013). Bioshock Infinite's forced baptism has sparked controversy and discussion. www.forums.ghostvolta.com/discussion/231/bioshock-infinites-forced-baptism-has-sparked-controversy-and-discussion/p1.

von Brauchitsch, M. (1943). *Kampf mit 500 PS* (4th ed.). Berlin: Karl Siegismund.

Walther, B. K. (2002, May). Playing and gaming: Reflections and classifications. *Game Studies: The International Journal of Computer Game Research*, 3(1). Retrieved May 13, 2004, from www.gamestudies.org/0301/walther.

Williams, B. (1988). *Upscaling downtown: Stalled gentrification in Washington D. C.* Ithaca: Cornell University Press.

Willke, H. (1987). *Systemtheorie* (2nd ed.). Stuttgart: Fischer.

Wilson, S. (2003). The aesthetics and practice of designing interactive computer events. Published in a different form in ACM SIGGRAPH 93 Visual Proceedings Art Show Catalog, 1993. Retrieved June 30, 2003, from http://userwww.sfsu.edu/swilson/papers/interactive2.html.

Winograd, T., & Flores, F. (1986). *Understanding computers and cognition*. Boston: Addison-Wesley.

Winter, R. (1995). *Der produktive Zuschauer*. Quintessenz, MMV Medizin: Medienaneignung als kultureller und ästhetischer Prozeß. Munich.

WoWWiki. (2013). The culling. Retrieved August 8, 2003, from www.wowwiki.com/The_Culling.

WuShogun212. (2013). Baptism controversy in Bioshock Infinite—man gets full refund. Retrieved August 8, 2013, from www.youtube.com/watch?v=Aa6nssPmQoE, My transcr., Apr 17, 2013.

Žižek, S. (1999). Is it possible to traverse the fantasy in cyberspace? In E. Wright & E. Wright (Eds.), *The Žižek Reader* (pp. 102–24). Malden: Blackwell Publishing.

Author Biography

Daniel Cermak-Sassenrath is Associate Professor at the IT University of Copenhagen (ITU), and member of the Center for Computer Games Research (game.itu.dk) and the Pervasive Interaction Technology Lab (PitLab, pitlab.itu.dk). Daniel writes, composes, codes, builds, performs and plays. He is interested in artistic, analytic, explorative, critical and subversive approaches to and practices of play. Discourses he is specifically interested in, are play and materiality, play and learning, and critical play. More information is available at www.dace.de.

Free of Charge

Julian Priest

Abstract *Free of Charge* is a participatory artwork that was first presented at the *Splore Music Festival* in New Zealand in 2012. It is staged as a mock airport security check procedure that is modified to measure visitors' static electrical charge. Participants pass through the security checkpoint and are measured for charge before being electrically grounded and discharged. The artwork and its site are described in detail and the rationale for the work developed. It is discussed in relation to the post-9/11 security apparatus and the concept of security theatre, and this is contrasted with aspects of the work that deal with health and wellness around static electricity. Through these lenses, response to authority and the internalisation and subversion of roles are examined.

1 Introduction

Free of Charge is a participatory artwork and performance first presented at the *Splore Music Festival* in New Zealand in 2012 (Splore 2011). It was commissioned for *Splore* by Letting Space (Letting Space 2013), a New Zealand arts organisation that specialises in contemporary artworks that are sited in public places outside of the confines of the white cube gallery space. The work was performed over the course of 3 days and had approximately 3000 participants.

2 Installation

Free of Charge consists of a temporary installation containing a mock airport security procedure similar to one found in any international airport (Priest 2014). The setup includes a roller table, possession trays, uniformed security guards and

J. Priest (✉)
Wellington, New Zealand
e-mail: julian@greenbench.org

© Springer Nature Singapore Pte Ltd. 2018
D. Cermak-Sassenrath (ed.), *Playful Disruption of Digital Media*,
Gaming Media and Social Effects, https://doi.org/10.1007/978-981-10-1891-6_16

Fig. 1 Free of charge: Being scanned

what appears to be a metal detector gateway. The gateway is topped with a LED sign displaying the scrolling message "Free of Charge" (Fig. 1).

For its first performance, *Free of Charge* was sited inside the *Splore Music Festival*, which is a bi-annual 15,000 visitor 3-day summer music, art and performance festival on the coast of the Hauraki gulf south of Auckland in New Zealand. Within the festival, the work was positioned on a grass area next to a track that linked two busy regions of the festival site, the main stage and one of the bar areas. The site was selected for its heavy foot traffic, and a natural constriction between a sloping hillside and the beach meant that large numbers of people passed by. Visitors were not forced to enter the work as it was placed to the side of the path rather than in the middle of it, and it was easily visible for several hundred metres in each direction. The work was sited on the foreshore and near to the beach, a position that is the de facto border of the country and the subject of recent political dispute in New Zealand over rights of ownership of the foreshore (Fig. 2).

Passers by had different reactions to the presence of a security check point inside a music festival, and this juxtaposition was intended to be visually jarring. Music festivals are presented as autonomous and free zones where the normal rules of society are suspended. People come to festivals to be allowed to transgress some of the rules of daily life—to live a hedonistic life outside of the normal constraints of work and leisure. The appearance of a security apparatus in this context was designed to highlight the reality of the music festival as a more tightly controlled space in which there are firm delineations between organisers, staff and festival goers. These are characterised by back stage passes and payment bracelets, and the entire festival is a walled zone with firm borders requiring payment to pass through the turnstiles.

A heavily policed border zone has been a feature of counter-cultural music festivals since the early days. At the 1969 Woodstock festival, 50,000 people entered without tickets before the fences had been secured, turning it into the iconic free festival of the time (Tiber 1989). Subsequent festivals were more careful to protect their economic interests.

In this context, the appearance of an additional security check, which appeared to be part of the real festival security provision, was accepted by most festival goers with many stopping to ask questions and begin an engagement with the work and only few being openly suspicious. The team of security guards included the artist, the curators and volunteers working in rotation in groups.[1]

When the work was empty, people would either ignore the piece or walk up to the guards and enquire what it was about. If there was even one person interacting with the piece, people were much more likely to engage and come over and ask questions about the work and often a queue would form (Aronson 2013). In this situation, people would begin to join the end of the queue without asking what the check was for, assuming it to be a mandatory checkpoint. As soon as the infrastructure was validated by one person's presence, other people would assume that it was official and join in. In effect, the validation of the group was responsible for bringing an infrastructure into existence, and in this way, the work models the social construction of infrastructural authority (Fig. 3).

3 Free

The work displayed an ambiguous scrolling message *Free of Charge* as an attractor. The title refers to the different meanings of free that have been endlessly debated in the discussions around free culture. The work offers an experience that is gratis and

---

[1]Guards roster included Julian Priest (artist), Sophie Jerram (curator), Mark Amery (curator), Trudy Lane (volunteer artist).

without cost, which was an attraction in the commercial environment of the music festival where food and alcohol were all charged via electronic cashless NFC wristbands. By contrast, the work's subject matter refers to the "libre" meaning of political freedom and liberty. Finally, the work was revealed as being a system to make the participant free of something, in this case both static electricity and role in relation to authority.

The initial user experience of the piece was to meet with a guard, find out what the security gate was for and then submit to their authority by removing shoes and emptying pockets. They would then wait to enter the scanner on a small carpeted micro-foyer area by the roller table.

The security guards were briefed to a loose script that evolved over the course of the work. The security guards were dressed in white shirts with epaulettes, aviator sunglasses and black cargo shorts but without shoes, in a parody of a security uniform. The dress code created an expectation from participants that the guards were authority figures, and people were willing to submit to their requests, "please take of your shoes and place them in the tray", "step up the plate please ma'am", "are you carrying any electronic devices? Place them in the tray please sir". This authority was also consciously undermined by informing people it was an artwork and the informality notes of the guards uniforms with no shoes and straw hats.

Most international holidays start with a security check so people are familiar with airport security procedures, and the form was easily appropriated and read. There is a way that people behave in relation to state border authority that is different from normal interpersonal relations and is a type of acting or role play. It is a moment in the theatrical performance of the state when citizens come into direct contact with the state's power of physical control. This moment of physical interaction is for most people a rare event that is not often approached in daily life. It can occur at moments of legal transgression or potential conflict. A driver being issued a speeding ticket, or a crowd being controlled at a football match for instance both come into direct contact with officers of the state. In crossing a border, we traverse the boundary of the state's jurisdiction and enter the suspended world of international transit. The air side areas of airports are highly regulated spaces with their own rules and regulations that are not always the same as their host countries. Liberties are reduced, authorities have extraordinary powers of stop and search and luxury shopping is encouraged in spaces that are perhaps models of an authoritarian consumer society.

Immediately after 9/11[2] security procedures around the world were strengthened dramatically for both national and international flights. Travelling just days before 9/11 on a New York to Los Angeles flight, I noted a sign on a hand baggage scanner saying, "Please remember to check guns in the hold". It is hard to remember a world in which people forgot to empty their pockets of hand guns frequently enough that a hand written reminder sign was necessary. Post-9/11 the

---

[2]11/9/2001 World Trade Center attacks, formatted here as 9/11 by convention.

airport security apparatus was heavily increased world wide and all security procedures tightened significantly. New technologies were introduced, and old ones steeped up so that now on most routes we experience not only bag X-ray checks, but also metal detectors, random frisking and pat downs, through clothes body scanners using both back scatter X-rays and microwaves, and chemical swabbing for plastic explosives (Health Physics Society et al. 2004). We remove clothes, belts and shoes, put toothpaste in plastic bags, surrender nail clippers and pen knives, empty water bottles and have *Nutella* jars confiscated to be consigned to the hazardous waste bins for immediate destruction.

4 Security Theatre

The practical and procedural measures of the massively expanded security infrastructure have made flying safer by thwarting attempts to smuggle weaponry on-board planes. Aside from the practicalities of security, there is a polemic and theatrical purpose which comes from the individuals' aesthetic experience of being processed and this is often referred to as security theatre (Schneier 2003). The term is normally associated with security measures that are designed to create a feeling of safety without actually being effective. Here, I do not want to focus on the effectiveness of the particular measures but on the experience of the performance itself as a theatrical and participatory work.

The theatrical performance of security infrastructure is designed to persuade the populous that it is safe to fly in times of asymmetric warfare. As co-opted improvisation actors in the security theatre, we are brought into contact with the security apparatus, feel the unfounded tension of being frisked and searched, feel the powerlessness of submission to the will of the state, feel guilt and uncertainty about an undeclared deodorant that hovers near the 100 ml limit and perhaps receive a reprimand for forgetting a pair of nail scissors. This theatrical performance with its roles of authoritative guard and submissive citizen reinforces for us our status and position within the apparatus. The security theatre creates an image of control that we internalise and carry with us in our daily lives, an image that precludes transgression, engenders compliance and helps to create the power relationships between individual and state.

The security theatre reminds us as audience of the existence of external threats and therefore helps to validate the rational arguments for the security apparatus and by extension the state as our protectors. In this way, the individual submission to the apparatus acts not only as an act of powerless submission but as entry point and membership of a powerful group. By acceding to the requests of the apparatus, we both enter into its protection and give it power by our tacit acceptance. The theatre is also designed to play to other audiences than the citizenry, to act as a deterrent and to demonstrate towards would-be attackers that there is a comprehensive security infrastructure in place.

Fig. 2 Free of charge: foreshore site

It is the conformance and acting out of these roles that continually reinforces and recreates the power relations. Every check is a dress rehearsal for the next check, an opportunity to hone the role of guard and citizen.

Similar roles were explored in the notorious 1971 Stanford Prison Experiment (Haney et al. 1973) in which a group of students was paid to take part in an experiment that modelled the prison environment. The students were split into two groups that were given the roles of guard and prisoner. The prisoners were arrested at their homes and incarcerated in a mock prison. Guards wore aviator sunglasses and uniforms, while prisoners were strip searched and de-loused. There was no explicit direction as to the behaviour expected of each group, and as the experiment proceeded, they begun to exhibit characteristics of archetypal guards and prisoners and to take on roles. The prisoners had a rebellion which caused the guards to respond with strong arm tactics and later on with psychological means such as solitary confinement and punishments. The experiment was designed to last for 2 weeks, but after 6 days, it had to be called off because the guards, and prisoners had inhabited their roles so thoroughly that the situation had become abusive and unsafe. Guards were regularly harassing the prisoners and one guard had become known as John Wayne for his strong arm tactics (Zimbardo 2014). After the early rebellion, prisoners entered into extremely submissive behaviour and begun to refer to themselves only by number rather than name echoing the TV show "Prisoner". The mock prison had become real.

The Stanford Prison Experiment demonstrates how ordinary people can readily internalise and adopt archetypal roles and exhibit extreme behaviours when placed in particular power positions within an infrastructure. In this sense, the power relations formed the roles, but the characters that they created for themselves were somehow known in advance by the students, learnt from direct contact with police and officials or learnt from the mass media, the heroes and villains of TV and film cop shows. The guards elevation to hero status and the prisoners' depersonalisation created an interpersonal gulf across which there was no basis for empathy and hence moral behaviour and cruelty was able to develop out of the group dynamic.

The roles in the security theatre are not arbitrary and do not stem from personal interpretation. In the case of airport security, they are contained in standards and classifications of people and objects that are negotiated and administrated by national border administrations and coordinated through international bodies such as the UN International Civil Aviation Organisation and the European Commission Directorate for Mobility and Transport. These organisations develop standards and offer training and assistance for their implementation, and the standards are adopted through the many bilateral air transport agreements. The standards and their underlying classification systems become boundary objects that inhabit both the communities of passengers and guards. In a process similar to that described by Bowker and Star, the standards become actors in their own right and push back into the biographies of their user communities, in turn setting up particular roles and behaviours in the individuals concerned (Bowker and Star 1999).

In response to 9/11, the ICAO Assembly understandably called for an increase in security through standards (ICAO 2014).

"[The Assembly] Urges all Contracting States to intensify their efforts in order to achieve the full implementation and enforcement of the multilateral conventions on aviation security, as well as of the ICAO Standards and Recommended Practices (SARPs) and Procedures relating to aviation security, to monitor such implementation, and to take within their territories appropriate additional security measures commensurate to the level of threat in order to prevent and eradicate terrorist acts involving civil aviation".

The implementation of the standards created the current infrastructural arrangements and that led to what I have termed the security theatre. Subsequently, these arrangements pushed back and began to determine behaviours that passengers exhibited and the roles that they internalised.

More than a decade on from 9/11, the security infrastructure has become so ubiquitous as to become invisible. In *Free of Charge*, a version of the security theatre was staged and its aesthetic experience used as the material for an artwork. The security infrastructure was brought back into view, and a setting was created within which participants' internalised roles in relation to security could be re-imagined and played with in order to encourage a re-evaluation, re-personalisation and re-humanisation of security procedures.

5 Electrostatics

While the *Free of Charge* scanner had the appearance of a metal detector, it actually housed a metal plate attached to the floor that was raised from the ground and electrically isolated on plastic stand-offs. Participants were asked to step up onto this plate one by one. The plate was connected to a sensitive electrometer, a scientific instrument that measures surface charge. It is similar to a standard multimeter, but has an extremely high input impedance, which means that charge does not flow through it when connected. This allows it to be able to measure the small amount of surface charge that is carried on objects, in this case on bodies. The surface charges carried on bodies sometimes have quite high voltages in the 1000 V + range, but only very small amounts of total charge. A normal multimeter will drain this charge in the process of trying to measure it in a fraction of a second and not be able to give a reading. Electrometers are typically used in industrial applications such as chip fabrication plants where static electricity carried on workers' bodies and clothes can damage sensitive electronics. The electrometer allows workers to see if they need to make themselves electrically neutral by grounding before commencing work.

Static electricity builds up on the body by friction of dry surfaces and insulators like plastic and synthetic fabrics, as well as by ambient charge from fields around electric installations. It persists in dry atmospheric conditions, and in urban locations, this voltage can become quite high 10,000 V or more. The common experience of a spark from a car door is caused by static and can only be felt if the static voltage difference is over 1500 V. In the festival setting with grass underfoot and the humid air near the beach, participants' readings rarely reached 100 V above ground.

Each participant was briefed about the possibility of carrying a static charge and had the measurement explained to them by a guard. Then, a voltage reading between the plate and ground was taken, triggered by a light sensor that the participant covered with their hand, and the reading was read out by a security guard.

Fig. 3 Free of charge: queue

6 Wellness

Having been alerted to this invisible but measurable charge, a short discussion evolved around what it meant for each person. In alternative health circles, there is a suggestion that excess static electricity is bad for you and that grounding can aid a number of complaints from sleeplessness to arthritic pain (Ober et al. 2010). There are numerous measures promoted to remove the static by grounding as an aid to wellness, from standing on the grass barefoot for 20 min a day to copper implants in shoes. There is no mainstream medical consensus about these practices or the dangers of carrying static electricity.

The role of the guards in this quasi-medical discussion was intentionally neutral. They moved from being authoritarian characters in a security apparatus to something akin to the role of doctor, health practitioner or counsellor. The conversation moved from being characterised by orders to listening and the tone softened. During the 2 min or so of engagement, a participant was shifted from security theatre to medical theatre, from one organ of the state, the security service, to the health service.

The emergence of the state as healthcare provider is described in Foucault's discussion of the UK post-1942 Beveridge plan (Foucault 2004). At a time when the state was engaged in large scale warfare, public health became a central object of concern. In many countries such as the UK, much of Europe and New Zealand, public health has been one of the great achievements of social policy, while adoption in the USA continues to be the object of hot political debate. Health has, however, continued to be a core justification for the state as well as in many cases one of it's largest expenditures.

"With the Beveridge plan, health was transformed into an object of State concern, not for the benefit of the State, but for the benefit of individuals. Man's right to maintain his body in good health became an object of State action. As a consequence, the terms of the problem were reversed: the concept of the healthy individual in the service of the State was replaced by that of the State in the service of the healthy individual".

While the medical establishment has been set up to support the health of individual, there is also an authority relation that develops, as well as numerous examples of classification and push back that have life and death impacts on the lives of citizens.

In *Free of Charge*, the health offering that was presented appeared to be from a scientific medical authority but was revealed to be an alternative medicine therapy. No strong claims were made as to the health benefits of the procedure, and each participant was left to draw their own conclusions. The medical authority in the piece was again left open to interpretation, discussion and play.

Each participant was then offered the opportunity to ground themselves by pressing a button with their toe which connected the plate to ground. As the button was pressed, any surface charge was able to flow to a large grounding rod buried in the earth and then people were able to see a zero reading on the meter. They let go of something, both the charge as well as perhaps any metaphorical baggage like a bad memory or emotion they wished to be free of.

The act of grounding signalled the end of their participation with the piece, and at this moment, they were free to go, to end their relationships to the various authorities in the piece, the artist and curator, the security guard and the medical advisor. Declared free of charge and grounded, they were free to walk barefoot onto the grass to reclaim their possessions from the tray and return to the de-limited freedoms of the festival.

7 Play

The piece set up a scene of authority that participants interacted with, but the scene was open to all kinds of interpretation by the public. People were aware that they were taking part in an artwork and that it was part of the festival programme, and this framing acted as an informal permission to subvert the structures. The piece developed into a rich performative world which encouraged subversion and playfulness, and there were many different interpretations of how to interact with it acted out over the course of the festival; one man staged dived the roller table sending everyone's shoes flying; another group used the static readings to decide who should buy the next round of drinks at the bar; a man measured his charge with and without a beach ball charged by rubbing it on his head; a group believed the scanner was a pregnancy test service; another pair walked through it with a bamboo poll carrying ten suitcases, and someone even fell asleep under the roller table. Many participants used the sand pit environment of the festival and the installation as an opportunity to playfully subvert their roles by experimenting with behaviour that would be considered as transgressive and potentially dangerous in live security screenings or medical consultations. The licence to live out playful reactions created a juxtaposition that brings into view the internalisation of roles in the real-life situations.

In participatory art, the artist creates a structure and encourages people to interact with it in particular ways. While appearing to be inclusive and open, there is often an aesthetic authoritarianism inherent in the form. Participants are stepped through prescribed series of actions in deference to the artist's vision in what can become an even more rigid aesthetic experience than for instance looking at a painting where the interpretation is open to the viewer. The work parodied this tendency with its mock security apparatus and the presence of both artist and curator dressed as guards and placed in roles of authority. The administrative hierarchy of festival, curator, artist, participant also mirrored the state hierarchy of state, government, civil service and public drawing a second parallel. The looseness of categories, shifting roles and festival setting worked to open a space for playful re-interpretation, subversion, individual responses and open conversation. As such the work can be read as a pretext for dialogue and the form encouraged a rich and personal series of discussions.

8   Conclusion

Free of Charge provided a safe temporary world which modelled the relationship of the individual to different forms of authority. The appropriation of the security theatre allowed the artwork to explore the way in which the security infrastructure comes into existence at the boundary between the formal structures of standards and procedures and the individual through the process of the internalisation of roles.

The juxtaposition inherent in a security check point for wellness invites questions about the role of the state in supporting or controlling the populous. The piece explored this by encouraging questioning, discussion, acting out of different behaviours and the playful subversion of the security infrastructure.

Stepping off the plate discharged, earthed and grounded, many whooped, felt elated, liberated, empowered, free of something intangible. They stepped away not only from micro-coulombs of static electricity, but also from images of the authority of state infrastructures, and the internalised roles formed in relation to them that we have come to take as our own.

References

Aronson, C. (2013). Love to queue (not). Retrieved October 2, 2014 from http://www.lettingspace.org.nz/free-of-charge-essay/.

Bowker, G. C., & Star, S. L. (1999). *Sorting things out: Classification and its consequences.* Cambridge, Mass: MIT Press.

Foucault, M. (2004). The crisis of medicine or the crisis of antimedicine. *Foucault studies,* No. 1 (pp. 5–19). http://rauli.cbs.dk/index.php/foucault-studies/article/view/562/607.

Haney, C., Banks, C., & Zimbardo, P. (1973). Interpersonal dynamics in a simulated prison. *International Journal of Criminology & Penology, 1*(1), 69–97.

Health Physics Society et al. (2004). *Public protection from nuclear, chemical, and biological terrorism: health physics society 2004.* summer school, Madison, Wis: Medical Physics Publishing.

ICAO. (2014). Doc 10022. *Assembly resolutions in force* (2013). Montreal: International Civil Aviation Organisation. Retrieved February 26, 2014 from http://www.icao.int/publications/Documents/10022_en.pdf.

Letting space. (2013). *Free of Charge.* Retrieved May 17, 2013 from http://www.lettingspace.org.nz/free-of-charge/.

Ober, C., Sinatra, S.T. & Zucker, M. (2010). *Earthing: The most important health discovery ever?* Laguna Beach, CA: Basic Health Publications.

Priest, J. (2014). *Free of Charge.* Retrieved February 26, 2014 from http://julianpriest.org/project/free-of-charge/.

Schneier, B. (2003). *Beyond fear: Thinking sensibly about security in an uncertain world.* Springer.

Splore. (2011). Splore—outdoor summer music and arts festival: Information. Retrieved November 13, 2011 from http://www.splore.net/#/information/getting.

Tiber, E. (1989). How woodstock happened. *Times Herald Record.* Retrieved February 12, 2014 from http://www.woodstockstory.com/how-woodstock-happened-1.html.

Zimbardo, P. G. (2014). The stanford prison experiment: A simulation study of the psychology of imprisonment. Retrieved February 21, 2014 from http://www.prisonexp.org/psychology/42.

Author Biography

Julian Priest is a New Zealand-based artist and writer who works with participatory and technological artforms. He was co-founder of the early wireless free network community Consume.net in London and is a board member of the Aotearoa Digital Arts Trust. He has lectured at the Banff Centre, Whanganui School of Design, AUT University and Massey University. Recent exhibitions include: The Blue Marble, Machine Wilderness, Public Art Finalist Exhibition, Albuquerque (2012); Sink, Machine Wilderness, ISEA, Albuquerque (2012–2013); and Local Time, Local Knowledge, Dowse, Wellington (2011–2012). His interactive sound work La Scala was recently commissioned for the Chartwell Stairwell at Artspace Auckland (2014–2015).

Playing on the Edge

Daniel Cermak-Sassenrath

Abstract Everything gets more interesting, challenging, or intense the closer it gets to the edge, and so does play. How edgy can play become and still be play? Based on Huizinga's notion of play, this chapter discusses how a wide range of playful activities pushes the boundaries of play in different and specific ways. For instance, gambling for money, party and drinking games, professional play and show sports, art installations, violent and military propaganda computer games, pervasive/mobile gaming, live-action role playing, festivals, performances, and games such as *Ghosting* and *Planking*. It is argued that in concert with a number of characteristics that mark an activity as play, play is essentially a subjective perspective and individual decision of the player. Huizinga calls this attitude the play spirit, which informs a player's actions and is in turn sustained by them. Edgy digital or mobile games do not challenge this position, but make it more obvious than traditional games that play is not only an activity, but a concept.

1 Introduction

In play, as in everything else, people push the boundaries to make it more thrilling, more intense, more fun. When Samuel Weber is struggling to define the "proper limits" between theater and ordinary life (Weber 2004, 314 qtd. in Crogan 2011, 139), he is pointing exactly towards the area many players and games are aiming for: play is pushed towards the everyday world.[1]

Here, several games, installations, and performances that challenge play are described and discussed: What are the boundaries in play? Which rules can be

---

[1]Nobody would be surprised if play were withdrawing in the opposite direction, to distance itself as much and in as many ways as possible from everyday life. But players are keen to go in the direction of everyday life. A play with play.

D. Cermak-Sassenrath (✉)
Center for Computer Games Research IT University of Copenhagen,
Copenhagen, Denmark
e-mail: dace@itu.dk

© Springer Nature Singapore Pte Ltd. 2018
D. Cermak-Sassenrath (ed.), *Playful Disruption of Digital Media*,
Gaming Media and Social Effects, https://doi.org/10.1007/978-981-10-1891-6_17

broken? How far can the boundaries of play be pushed? Is it still play, then? Who decides? What can one learn about play from this?

The discussion is based on the notion of play proposed by Huizinga. Play is seen as an attitude of the player that expresses itself in a process that is marked by multiple characteristics. Huizinga (1955, 132) describes play as

> [A]n activity which proceeds within certain limits of time and space, in a visible order, according to rules freely accepted, and outside the sphere of necessity or material utility. The play-mood is one of rapture and enthusiasm, and is sacred or festive in accordance with the occasion. A feeling of exaltation and tension accompanies the action, mirth and relaxation follow.

The perspective and the process of play interact and complement each other.

It appears that the material, physical, tangible, or visible activity is almost arbitrary, as long as it is informed by the playful attitude of the players. Players make the decision to play individually; they take on and share this special attitude. They turn something into play. They create play for themselves from (nearly) everything and everywhere at any time. With digital media this becomes more visible than before.

2 Characteristics of Play

Play *is* not (in the sense of an artifact or product) but is *being made to happen* (in the sense of a process). According to Huizinga, the process of play is formally marked by multiple features: play is free, is without end and can be repeated, requires and produces order, is marked by a certain tension, is distinct from ordinary life, is not connected with material interests, and deeply immerses the player.[2]

The German pedagogue Hans Scheuerl defines play very similarly and adds the characteristic of *Scheinhaftigkeit*: play oscillates between the poles of everyday life and illusion (1965, 83) without ever reaching one of them. He draws on Schiller's notion of the aesthetic appearance [*Schein*] in discrimination from the logical appearance: the appearance "that we love because it is appearance" (ibid., 84, my transl.), and not because we are fooled.

These characteristics interact, build and depend on each other, and together form an integrated unit. They appear to Scheuerl as "only different ways in which the same phenomenon is represented" (ibid.:79, my transl.): play.

If and as long as these features of play characterize an activity, it can be play for a player. But there is no automatism. He or she verifies their continued existence, and

> is constantly noticing if the conditions for playing the game are still being met, continuously monitoring the "frame," the circumstances surrounding play, to determine that the game is still in progress, always aware (if only unconsciously) that the other participants are acting as if the game is "on." (Sniderman, 2 qtd. in Salen and Zimmerman 2004, 94)

Players decide individually if they will keep on playing, and the question of whether someone is playing can only be answered by that person.

---

[2]For a recent look at Huizinga's notion of play in the light of digital media see Valerie et al. (2015).

3 Play as Perspective

Following Huizinga, one cannot only ask for "purely formal" characteristics of play, but for the "attitude and mood of play" (1955, 20), how players themselves experience play.[3]

The playfulness of a game depends on a specific attitude of the players (Scheuerl 1965, 106). To play means to act from a certain perspective. This perspective is the play spirit, a mood that players willingly take on and that simultaneously captivates them. "In play as we conceive it the distinction between belief and make-belief breaks down" (Huizinga 1955, 25). For Buytendijk, "playing is always playing with something, that also plays with the player" (Buytendijk 1933, 161 qtd. in Retter 2003, 16, my transl.). Acting in the play-mood is simultaneously pushing and being pushed; being in complete control and giving up all control; keeping up play and being swept away by it: "Whether one is sorcerer or sorcerized one is always knower and dupe at once. But one chooses to be the dupe" (Huizinga 1955, 23).

Play is hovering above ordinary life (Fischer 1925, 69 in Scheuerl 1965, 81), and to move in the world of play means to

[D]well in the realms of chivalry and heroism, where illustrious names and coats of arms and splendid lineages bulk large. This is not the ordinary world of toil and care, the calculation of advantage or the acquisition of useful goods. Aspiration here turns to the esteem of the group, a higher rank, marks of superiority (Huizinga 1955, 60).

Players "dare," "take risks," "bear uncertainty," and "endure tension"; "these are the essence of the play spirit," writes Huizinga (ibid.:51). Bernard Suit's "lusory attitude" is a

[S]tate of mind whereby game players consciously take on the challenges and obstacles of a game in order to experience the play of the game itself. Accepting the artificial authority of the magic circle, submitting behavior to the constraints of rules in order to experience the free movement of play, is a paradoxical state of mind. (Salen and Zimmerman 2004, 574)

For Bateson, the play attitude is a "delimited psychological frame, a special and temporal bounding of a set of interactive messages" (Bateson 1971, 191 qtd. in Salen and Zimmerman 2004, 370). There are other notions that closely relate to this idea, such as Eugen Fink's "*Spielsinn*" (Fink 1957 qtd. in Retter 2003, 37).

Play is not only an activity, but an idea. The activity does not create play but articulates the play spirit. Playing a game is transforming an attitude into a process. Buytendijk observes a "unity of mood and movement" (Buytendijk 1933, 59 qtd. in Retter 2003, 16, my transl.). Activities are always informed by the perspectives of those who perform them (Dourish 2001, 204). Pure activity is not interesting for play and is not even play, because "for whatever [...] play is, it is not matter" (Huizinga 1955, 3). Play is only interested in the "ideal fact that the game is a success or has been successfully concluded" and not in the concrete representation or in "the material result of the play, not [in] the mere fact that the ball is in the hole" (ibid.:49).

---

[3]This reading of Huizinga has been presented and discussed e.g. in Cermak-Sassenrath 2010a, 178ff., Cermak-Sassenrath 2010b, 83ff. and Cermak-Sassenrath 2015, 94ff.

The visible action appears as the result, expression, and reflection of the primary perspective. The play action in itself appears trivial and useless. Only seen from inside play it acquires meaning, value, and sense: "The act of play is the act of interpretation" (Salen and Zimmerman 2004, 372) (see the chapter "Makin' Cake-Provocation, Self-Confrontation, and the Opacity of Play" in this volume).

Play is only play when it is experienced by somebody (Scheuerl 1965, 192). People perceive play in different activities, and play is not limited to certain activities. Play is opaque with regard to the meaning of actions and objects in games (Cermak-Sassenrath 2013). Media always have content, but the content is arbitrary and exchangeable. There appear to be few topics or genres that not lend themselves to play; borders are drawn by tradition, preference, and taste.

> Could you make an entire game out of an experience that is typically ordinary or tedious? How about a game designed to be played while waiting in line? Or watching the news? Or driving a car? Once you understand that play is latent in any human activity, you can find inspiration for play behaviors and contexts anywhere (Salen and Zimmerman 2004, 306–307).

Play is not limited to casual, easy, or low-risk activities. "The Japanese samurai held the view that what was serious for the common man was but a game for the valiant." (Huizinga 1955, 102)

It is a conscious decision to play. It is not entering a tennis court, wearing a jersey, or holding a golf club that makes somebody a player, but the mental change from everyday life into the play world. This "*stepping out of* common reality into a higher order" (ibid.:13) is the precondition and effect of play.

Within play, players need to make rational and efficient decisions. But the decision to play is not rational. Being rational is not fun. Huizinga rejects explanations that reduce play to a secondary (e.g., biological or social) phenomenon. For him, *there are no* secret or hidden motivations, reasons, or purposes in play to be uncovered:

> ... the *fun* of playing ... resists all analysis, all logical interpretation. As a concept, it cannot be reduced to any other mental category. ... It is precisely this fun-element that characterises the essence of play. Here we have to do with an absolute primary category of life, familiar to everybody at a glance right down to the animal level. (ibid., 3)

4 Games that Challenge Play

Games move towards ordinary life by reducing the distance and "undermining the legitimacy of [the] separation" between their "virtual suspension of the real" and "the serious business and high stakes of real-life conflict" (Crogan 2011, 140). Here, several examples demonstrate how and in which ways many traditional games as well as digital games challenge play and push the boundaries of what is allowed/expected/accepted: *Ist erlaubt, was gefällt* (Wilhelm Busch)?[4]

---

[4]Is [everything] permitted that pleases?

Connections to Huizinga's theory of play are established, and it is discussed if the magic circle of play is indeed in danger. A categorization is proposed; there appear to be certain types of games that question a particular aspect of play, or play off a particular trespass.[5]

Many or even the majority of the games discussed here are influenced or even driven to a considerable degree by social interactions between players and/or between players and an audience. These processes are not phenomena specific to games, and are discussed here only if they are the focal point of the particular aspect that a game challenges (e.g., in party games).[6]

4.1 Games of Chance

There are heaps of games in which the players' success depends to a significant degree on luck or chance. Games in which players gamble for money (e.g., roulette) are prime examples. Relevant here is the moment of random chance these games heavily (or even exclusively) rely on, not the material (e.g., monetary) implications of playing them (which are discussed in Sect. 4.3).

Although sometimes random chance is seen as an essential property of play (e.g., by Eigen and Winkler 1990, 11 qtd. in Keller 1998, 237), what Salen and Zimmerman term *meaningful play* can only happen when players are in control and have (exclusive) control over the course and outcome of a game.[7] In skill-based competitive games, random chance is routinely reduced by players, for instance, by switching off the "questions round" between races in *Super Cars 2* in which players can gain (but mainly lose) championship points and money. The argument that games of luck are not genuine games is that there can be no freedom of decision in a game that is random or, rather, there can be no freedom of decision when there is nothing to decide on, or the decision has no bearing on the outcome of the game (Fig. 1). Play and chance seem to contradict each other conceptually.

One can say that chance inhibits play. But judging by everyday experience, chance is no obstacle for play. Many playful activities involve a high degree of chance, and can hardly be called anything other than games.

There are two ways to explain why these activities are called games, and indeed can be games. One is that players play with randomness, take on chance itself, and believe they can win (Keller 1998, 60f.). If gambling is random chance, probably it can be beaten. Such a player tries to overcome fate itself. "He treats something he cannot control as though he could control it. At one moment he thinks it is in his hand, in the next moment it slips through his fingers." (Scheuerl 1965, 153, my transl.). Scheuerl calls this type of player a "gambler" [*Hasardeur*]. The other way is

---

[5] Where a game touches upon several aspects, one of them (probably the most obvious, interesting, or unusual one) is chosen.

[6] The role of the audience in games such as *Tekken Torture Tournament* and *PainStation* is discussed, for instance, in Crogan (2011), 136–138.

[7] Random chance is not identical with play's ambivalence.

Fig. 1 You will win! ad
(Magnamail Pty Ltd,
Auckland, catalog 2104 NZ,
n.d.)

when people do not believe in chance. If there is a system governing what happens, it must be possible to decode and understand it (Keller 1998, 60f.)—then luck is a skill that can be mastered.

In both cases the playfulness of the activity is not questioned. "*Skill-based games and games of random chance lie on the same line*: In one case an opponent is to be beaten, in the other chance or fate itself" (Scheuerl 1965, 153, my transl.).[8] Claude Shannon (qtd. in Roch 2009, 164) describes play as a three-polar system between "Chance, Physical Skills, and Mental Skills." Players play and win games by skill or luck. They accept a challenge they believe they can master or overcome.

Players need control over their play. But for a game to engage, fascinate, and thrill, to take off and fly, it needs to get out of each player's control; it needs to get out of hand in a way. Play hovers between being completely controlled and having totally unpredictable behavior. In both of these extreme cases it loses appeal and is not (or no longer) played. This is the same for both games of skill and games of luck. Gambling maximizes the aspect of play slipping from players' control. Nobody knows how it is going to end, and nobody *can know*. But this increase in thrill goes at the expense of meaningful action.

Play is always open-ended; this is nothing that is connected to gambling. It has to be, and nobody within the game (i.e., players, teams) has to have complete control over it, and nobody without has to have it, either: All players can potentially win their games. If they do not see this possibility or at least a slim chance, they do not play. If a game is decided (or rigged) it does not thrill (anymore).

Games of chance function as games as long as they follow certain rules and move inside certain borders, are restrained, regulated, and not completely arbitrary. Play is perverted when players are deceived about their possibilities of influencing the

---

[8]This seems to be similar to how some people see ordinary life.

course and outcome of a game (see e.g., the chapter, "Programming Chance: The Calculation of Enchantment," in Schüll 2012, 76ff.). This happens for instance in gambling machines such as one-armed bandits: players are asked to act and misled to assume that their actions are important, but their decisions do not change the outcome in any essential or significant way (Keller 1998, 60f.).

4.2 *Strip Poker, Kissing Games, and* Truth or Dare

Some games purposely challenge social norms or conventions. There certainly appears to be a special thrill to these games. The attraction lies not only in playing them, but also in openly defying and rejecting accepted rules. But playing such a game also means to leave behind usual securities such as etiquette, good taste, and the like.

Twister is a round-based game in which several players try to reach and occupy certain spots on a mat with their hands and feet. People have to keep their balance in weird and often awkward positions. Touching other players is unavoidable, and adds a considerable social thrill to an otherwise trivial game. The manual (Hasbro 2012) is aware of this, and at least suggestive. Its second sentence reads: "Twister can be played indoors or outdoors by boys or girls or mixed groups of all ages." Unsurprisingly, in the mid-1960s, when the game was first released in the United States, it created some controversy (About.com 2012). *Twister* became popular after "Eva Gabor played it with Johnny Carson on television's Tonight Show on May 3, 1966" (Wikipedia 2012b).

Where *Twister* invites full body contact between players, the *iPad* game *Fingle* (2011, Game Oven, fingleforipad.com, Fig. 2f.) focuses on hands and fingers.

> Fingle is a cooperative two-player iPad® game about the thrills of touching each other on a multitouch device. Two players drag up to five buttons of one color onto their matching targets; their movement makes it impossible to avoid contact, creating intimate moments with intertwined hands (Game Oven).

The music of the game might or might not resemble soundtracks of 1960s porn movies (Fig. 3).

Fig. 2 Players at *Fingle*
(*image* Game Oven 2011)

Fig. 3 *Fingle*: "Pick me up," "hard to get" (*images* Game Oven 2011)

Fig. 4 *Shadow Showdown*:
Team play (Creativity &
Cognition 2013 conference,
Sydney, Australia)

Shadow Showdown (Martin et al. 2013) (Fig. 4) invites whole-body interaction between players. They use their bodies to mimic and fill in specific shapes displayed on a screen. After 15 s, a snapshot is taken and correct silhouette coverage is awarded percentage points. Players can either participate in teams or alone, cooperatively or competitively. The game plays off players' bodily movement, proximity and touch, as many levels require players to stand/duck/crouch close to each other, to hold or even carry each other.

Truth or Dare is a classic party game: Players

> take turns either answering a "truth" question, or performing a "dare." …Games can range
> from funny to serious, from intimate to wild. …The key to a good game of truth or dare is
> in the availability of good truth questions, fun dares, and a like minded group of people who
> are uninhibited and ready to have a great time (WSRA, Inc. 2012a)

As with many or all party games, this is a game without spectators—everybody present needs to join the game or leave.

An example truth question is: "If you were homosexual which of your friend[s] would you find the hottest?" (www.truthordare.us (May 7, 2012)). An example dare is: "Draw a mustache with a marker or pen on your face" (www.getdare.com (May 7, 2012)).[9]

---

[9]Truth questions and dares are readily available online, including disclaimers, for instance:

> Always take proper safety precautions, and NEVER perform any dare that is illegal or would
> result in physical or mental harm to anyone else. By using this site you warrant that you are
> of age to view the content presented, and accept full responsibility for any actions arising

Fig. 5 *Broken Picture Telephone* example

Fig. 6 Playing *Cards Against Humanity*

Broken Picture Telephone is a *Chinese Whispers* variant: something is alternatingly drawn and described in writing by the players (Fig. 5). The game is collaborative and premised on the understanding that all players do their best to transmit the message. It is a social game without a winner. A lot of the playing time is usually spent speculating, discussing, and explaining what is (and was) going on (before). The thrill is the interplay of the game mechanic with the act of interpretation and association: weird, serious, touchy, sophisticated, and funny aspects of life are seen through 30 s drawings and explanations. Playing the game involves recognizing and utilizing people's real world knowledge and cultural backgrounds.

Cards Against Humanity is a multiplayer game (Fig. 6), organized in rounds. In every round one player asks the other players a question or reads a short cloze text, and the other players select one or several of the answer cards they hold and covertly hand them over to the asking player. He/she then reads the answers aloud and selects the one which He/she likes most, for example, finds the funniest. The player whose answer was chosen gets a point. On the game's website[10] myriad user-made question and answer cards are available. They range from harmless to politically totally inappropriate—topics include minorities, sex, body functions, violence, celebrities,

---

from this site's use both directly or indirectly. … All content is for entertainment only. Keep it safe and keep it fun! (WSRA, Inc. 2012b).

[10]https://cardsagainsthumanity.com/.

history, children, and a mix of those. The game appears to be darker and edgier than the similar *Broken Picture Telephone*.

Like many games, seen rationally, this game is quite trivial. The thrill appears to be entirely drawn from the clash of the safe and distinct place inside the game's magic circle with the hilarious connections, associations, and references to ordinary life. Players select from quite limited and often weird or dodgy options, and they submit their answers anonymously. Then the player who asked the question reads out loud what players submitted; only the player who submitted the winning entry is then asked to identify him/herself. The winning mechanism is highly subjective and intransparent (although fair because the entries are anonymous); it is not a very competitive game but a social and communicative one.

In the dice and drinking game *Mia* (German *Meiern*), players need to roll increasingly high numbers, or successfully convince the next player that they did. Winning players are either lucky or have perfected their skills of lying.

A specific boundary that several of the games described above push, is morals. It is easy and not risky for play, because morality is not a category of play. "Play [...] lies outside the reasonableness of practical life; has nothing to do with necessity or utility, duty or truth" (Huizinga 1955, 158). The notion of morals is not a notion of play, but of ordinary life. "Play [...] lies outside morals. In itself it is neither good nor bad" (ibid.:213). Schaller (1861, 85 qtd. in Scheuerl 1965, 117, my transl.) notes that "[i]t is neither a person's duty to play nor to not play, it is permitted. It is neither virtue nor selfishness. It is innocent" (cf. ibid.:78). Inside play, there are only actions that have abstract meanings, that have meaning only inside play, and only players can know "what things mean" (Salen and Zimmerman 2004, 366). Play cannot be judged either by outside rules, norms, and conventions or by morals or laws (Krämer 1995 qtd. in Pias 2002, 158) of everyday life. If this were the case, many games (e.g., soccer) would appear to be ruled by the laws of the jungle, that is, skill, force, or luck. For Schiller, play allows people to take on a "third character," in which they are free from compulsive desires [*triebhafter Begierde*] as well as from moral constraints (Scheuerl 1965, 70) (See the chapter, "Makin' Cake-Provocation, Self-Confrontation, and the Opacity of Play," in this volume.).

The games described above challenge the characteristics of freedom, dividedness (as ordinary life is influenced by these games, which relish in trespassing the boundary of play), and immediacy of play (players have to stay in-play, regardless of the actions that are out of the ordinary).

It is not that these games allow everyday life to intrude into play: it is play trespassing on the ordinary world. One can call this medial spill or bleed, as it happens in other media as well:

> Hollywood is a town of fabulators. The people who dwell there create fictions for a living, fictions that refuse tidily to confine themselves to the screen, but spill over into the daily lives of the men and women who regard themselves as stars in the movies of their own lives (Biskind 1999, 7).

In strip poker and *Truth or Dare* the challenge for players appears to be to play the abstract game successfully and simultaneously to be aware of the sensual/visual/concrete thrill; the skill that is required is to handle this special duality, and the continuous changes from one state to the other. Players have to be aware of the fact that their actions are solely abstract in the game, and, on the other hand, concrete in the ordinary world. They do not decide, but go both ways, and balance between play and everyday life. Other than that, these games are trivial, and very robust. These games also require security (discussed in Sect. 4.5).

Games can effortlessly break many rules and conventions and show how arbitrary and vulnerable they are. They do not care about morals; and players are not (or no longer) bound by artificial and arbitrary ideas about ideal life. This is fun, and it is even more fun to trespass on the rules that have been left behind. Players feel empowered by this act of liberation through their playing. At the same time games introduce their own rules, and sweep players off their feet, as they realize the usual rules and customs no longer count.

4.3 Professional Play

Throughout history people have been paid to play, for instance, musicians, circus artists, and athletes. Today there are professional *Starcraft* players and *World of Warcraft* farmers. Dennis Rodman says of his playing professional basketball in the NBA: "Fifty percent of life in the N.B.A. is sex. The other fifty percent is money" (Dennis Rodman qtd. in Brainy Quote 2012). The installation *Choco-Pacman*[11] (Fig. 7) gives out chocolate to (winning) players. The official song of the Olympic Games 1988 in Seoul "Go for Gold" (written by Wolfgang Jass) suggests that winning (Olympic) games was about "gold" and "treasure":

> Go for gold in South Korea
> Go for gold in'88
> There's a pinnacle for climbing
> Down in Seoul the treasure waits (fritz51338 2012, magistrix 2014)

But also every increase in popularity between friends and in the level of bodily fitness is an effect of play in the everyday world that might interact with playing.

Games that have pronounced effects beyond themselves, such as gambling, are sometimes located outside genuine play (e.g., by Jesper Juul 2005, 43) because their "real-world stakes ... constantly threaten to destroy the nonserious, playful quality thought to be a core characteristic of genuine game-based activity" (Crogan 2011, 30). The first characteristic of play endangered by material implications is its purposelessness.

---

[11] *Choco-Pacman* by Malgorzata Dabrowska and Florian Keßeler, developed in Daniel Cermak-Sassenrath, Bernard Robben, Susanne Grabowski. Art in Action: Computerspiele, and interaktive Kunst und neue Schnittstellen (Computer Games, Interactive Art and New Interfaces), Course, University of Bremen and University of the Arts (Hochschule für Künste), Bremen, Winter Semester 2009/10.

Fig. 7 *Choco-Pacman*
installation (*image* Bernard
Robben 2010)

Artists are paid for their work and they create art; this is hardly questioned or even discussed,[12] whereas professional athletes are suspected of not playing or at least not really playing.

Effects of play in ordinary life do indeed challenge play. But even then, play remains a decision of the players, an irrational position. People play in expectation and acceptance, and in open defiance. In many cases, it appears, play's immediacy and players' intense focus on play precludes any consideration as to what effects may result: players are thinking about winning, not money.[13] That is to say: play does not care about its setting at all (Retter 2003, 22, 144).

Occasionally, professional soccer players cry after losing a match, for instance, the players of Portugal after they lost the EURO 2004 final to Greece, or the players of England after losing to Portugal at the 2006 FIFA World Cup quarter finals.[14] Observations such as these support the statements of professional athletes who often explain how dominant they experience play. Boris Becker answers the question if professional sports "is all about money," by saying that "[i]n the first place, it is about play" ("*Es geht in erster Linie ums Spiel.*") (Boris Becker in RTL 2005, my transl.). Dirk Nowitzki states that money is not a motivation for him; he says he plays because he "love[s] basketball, want[s] to compete with the best and win the [*NBA*] championship [2006]" (Dirk Nowitzki qtd. in Schlickmann 2006, 69, my transl.). *Formula 1* race-car driver Nick Heidfeld asserts that "fun was and continues to be the only reason [he is] in it at all" (Heidfeld 2007, before the Grand Prix of Belgium at Spa-Francorchamps, my transl.). For John McEnroe, professional sportsperson is "one of the best jobs there is. And you can't really call it a job either" because it is so much fun (John McEnroe qtd. in Wichert 2006, 74, my transl.). Motorcycle racer Colin Edwards states that at 38 years of age his career is likely to be over soon. But if somebody offered him a possibility to keep on racing "for another ten years," he

---

[12]Several works by Koblin address the issue of paid distributed online labor in an art context (e.g., *The Sheep Market* by Aaron Koblin (2006, Fig. 8), *Ten Thousand Cents* by Aaron Koblin, Takashi Kawashima (2008, Fig. 9)).

[13]Analogue to Brecht's *Erst kommt das Fressen, dann kommt die Moral* (literally, "First comes food, then morals.").

[14]Presumably they were not crying because they lost the prize money (if there was any).

Fig. 8 Tool for drawing
sheep for *The Sheep Market*
(*image* Aaron Koblin 2006)

Fig. 9 *Ten Thousand Cents*,
website of finished work,
replay of the drawing of a
part (*image* Aaron Koblin,
Takashi Kawashima 2008)

"would accept immediately." He says he had originally planned to stop professional
racing at 32, but when he reached that age, he thought, "This is way too much fun"
(Colin Edwards qtd. in Scott 2012, 38, my transl.).

4.4 Play as Show

Playful activities such as games and sports draw huge crowds. Fans follow events
and stars enthusiastically, and with great investments of energy, time, and money:
and they are far from being passive (see, e.g., Winter 1995). Spectator sports are
also a large commercial market (e.g., soccer, Olympic Games, *Formula One*). Play
has always been used or misused for purposes beyond itself. The delegation of play,
and the use of play as a means to appease people, to keep them from engaging in the
political arena and to entertain them (Christ 2005, 789) happened on a large scale
in the *imperium romanum* since Caesar. It was excessively done by Trajan, who in
the period of three years initiated 147 days of fighting to celebrate the victory in
the second Daker war, involving 11,000 wild beasts and 4914 pairs(!) of gladiators
(ibid.:300). A participant in such an event is what Philip Roth (2002, 41) calls a
"surrogate" for the spectator, a "stand-in," "in [the] service" of him, who becomes

"an invisible accomplice in the act." The idea that play's purpose and role is entertainment, is explicitly formulated by Chamblanc, for example,"We must not swerve from the main purpose that a game is meant to entertain" (Chamblanc 1828, iv qtd. in Pias 2002, 173, my transl.).

It is nothing new today if play is proposed and used as a means of entertainment. There is a clear tendency for competitive sports to move towards providing a good show for an audience that is often propagated quite openly and positively, for example, in motor sports: "Car racing is a show and always has been one. It is about entertainment and emotion and nothing else" (Lacroix 2012, 24, my transl.). Cynically, *MotoGP* driver Jorge Lorenzo suggests rewarding the impressively risky driving style of Marc Márquez, which he considers to reward for the other drivers and for the track marshals, with additional championship points. He says that it was "a good show" when Márquez pushed him off the track in Jerez, crashed under yellow flags in Silverstone, and ripped off Dani Pedrosa's traction control cable while trying to overtake him in Aragon (Jorge Lorenzo qtd. in Motorsport aktuell 2013). The characteristic of purposelessness is challenged when play is used instrumentally as a means to entertain rather than performed for its own sake and the sole benefit and exclusive enjoyment of its participants. It appears that many players cannot afford to voice their (supposedly critical) views because they depend on sponsors' or public support.

In TV broadcasts of sport events, entertainment increasingly appears to push back sports. Playing becomes only the background activity for lotteries and interviews with stars (Deutschlandfunk 2006). It is commonplace to say that if something is not attractive to look at, nobody wants to see it or to pay for it. Events are scheduled at times of day for convenient live TV coverage (e.g., night races in *Formula One* in Singapore and in the *MotoGP* in Qatar; the NZ Super Rugby League starts matches at 7:45 PM for live TV transmission to the United Kingdom and South Africa).

But people argue also for "protecting the central aspect of sports [women's 200-m freestyle swimming, beach volleyball, etc.] from being overloaded with entertainment" (Gunter Gebauer in Deutschlandfunk 2004) because competitive games need to remain play for its players, not only shows for spectators. Play does not depend on spectators, who are optional: "My passion comes from my love of the game. ... It doesn't matter if there's 100,000 fans in the San Paolo in Naples, or 2,000 fans here, it doesn't matter how many fans there are, I'm always passionate about this game" (Diego Maradona qtd. in O'Brien 2012). Competitors "wrestle in an empty space, to an audience of none" (Irving 2006, 292). Athletes are not performers as in the theatre; they are (game) players.

Sports is *not only* entertainment (Gunter Gebauer in Deutschlandfunk 2004, my emph.), and play not a means to create a spectacle for an audience: "[V]ideo games ... may turn out to be far more than entertainment" (Norman 2004, 133–134). As an attachment to entertainment play is endangered. It is not disputed here that play can entertain spectators, and in a certain sense also its players; it can also train bodies and improve social skills, among other things; but these play-external effects do not describe or even touch its kernel or essence. For Jean Château, "the delight or joy of playing is not born in simple amusement" (Château 1946, 31; Scheuerl 1965, 111, my transl.). Play entertains as a secondary effect, incidentally. Play is not aiming to fulfill any purpose beyond enjoyment in itself for its players.

4.5 Risky and Violent Games

All games play for something and are risky; many involve bodily movement and action, and some (digital and analog games) are very violent. Including real-life effects such as bodily injury in play increases the wager that is played for, similar to gambling for money. Of course, this challenges play, as does involving money, but used moderately it might increase the thrill of play: "Obviously you get more out of the game when you're playing with higher stakes, and in motor racing you're playing with the highest stakes of all" (Moss 1959, 6). Games such as the *PainStation* (Fig. 10) challenge a number of characteristics of play: the dividedness of play from everyday life is challenged by the bodily injury that might result from playing; the very high tension connected to playing these games challenges the ambivalence of play; if there is no smooth integration of play actions and play effects within the games, the appearance and the immediacy of play are challenged.

All kinds of games can involve consequences that spill out into everyday life as part of playing: children's games, professional sports, party games, and so on. There are many examples of games in which players take considerable risks. In some games, the possibility of (sometimes severe) bodily injury is integrated or identical with the play activity; in some games bodily punishment is deliberately added on to an otherwise low-risk play activity.

In the (noncomputer) game *Lightning Reaction* (2005) a red light goes on, all players press a button on their sets, and the last player to do so gets an electric shock (Alexander 2005). In *Shocking Tanks* (Fig. 11) players control RC toy tanks and shoot at each other's tanks; the player whose tank receives a direct (infrared) hit gets shocked (ibid.). A spiced-up version of arm wrestling ("The ultimate in hand to hand combat" (package slogan)) is *Shocking Arm Wrestling* in which the loser receives an electric shock. In the movie *Never Say Never Again* (1983) James Bond

Fig. 10 *PainStation* players (*image* //////////fur//// art entertainment interfaces 2003)

Fig. 11 *Shocking Tanks*

is playing *Domination* against Ernst Stavro Blofeld which involves electric shocks of potentially lethal strength for the losing player. There are card games in which players are punished in various ways after they lose a round.

Car and motorbike races involve crashes; they require drivers to put themselves on the line, although racing could easily be realized with simulators, without any dangers for competitors or spectators, and no ill effects to the environment. In boxing, opponents punch each other and knock each other down and even unconscious. In *5-finger fillet* (Fig. 12), "[A] person places the palm of his or her hand down on a table with fingers apart, [and] using a knife, or sharp object, ... attempts to stab back and forth between [his or her] fingers, moving the object back and forth, trying to not hit [them]" (Wikipedia 2012a). Drinking games integrate play with the consumption of alcoholic drinks and players getting intoxicated.

There also have been several attempts to increase the thrill of digital games with special controllers. For instance, the US-based company Mad Catz developed a gamepad called the *Bioforce* that is able to give the player 16-milliamp electric shocks as feedback, for example, in fighting games. A prototype was shown at the Electronic Entertainment Expo in Los Angeles in 2001, and the commercial release was scheduled for late 2001 or early 2002 (McCarthy 2001; Marriott 2001; Ulie 2001). It appears, though, as if it were never released.

Fig. 12 *5-finger fillet* player

Fig. 13 *Tekken Torture Tournament* (apparently at Adelaide Film Festival, Adelaide, Australia, Mar 4, 2003) (*image* Eddo Stern, Mark Allen 2001)

Artistic installations take up the issue of play and violence. *Punch It*[15] seems to be a simple punching game in which the player tries to knock out a target indicated by a spotlight as fast as he can, without hesitation. But the targets are humans with large pushbuttons strapped to their bellies.

The *PainStation* is a version of the popular arcade game *Pong*. Two players face each other across a horizontal screen. The paddles are controlled with the players' right hands, with rotary knobs, while their left hands have to hold on to metal contact plates, called the Pain Execution Units (PEUs). The original game play of *Pong* is unchanged, but an essential modification is that if a player misses the ball, she receives unpleasant feedback through the PEU: a mild electric shock, a short wire whip, or a brief heat impulse. The winning player is not determined by a score; the *PainStation* is not so much a game of skill playing *Pong*, but players test their abilities to withstand bodily abuse: the player who retreats loses.

A well-known installation using the *Playstation* game *Tekken 3* is C-level's *Tekken Torture Tournament* (2001). In events held around the world (Fig. 13f.), "[w]illing participants are wired into a custom fighting system ... which converts virtual on screen damage into bracing, non-lethal, electric shocks" (www.c-level. org/tekken1.html (Feb 3, 2012)) (Fig. 14).

In artistic performances, artists such as Mike Parr and Orlan (Crogan 2011, 202, footnote 7), Marina Abramović, Chris Burden, and Stelarc play with the audience, question, and push the boundaries of morals and ethics, of what spectators expect and tolerate, and are willing (or enjoying) to endure.

The performances by Andy Kaufman appeared to be genuine and seemed to leave the fictional level of entertainment, as he was confronted, threatened, beaten up, and the like, but they were not. Incidents such as the famous brawl on a live TV show (Fridays, ABC, 1981) were planned. Nobody, least of all Kaufman himself, explicitly claimed they indeed were real, but many spectators felt cheated. Kaufman probably also faked his death in 1984.

But it does not need to be a performance in which somebody deliberately harms herself. In many performances risk is a part of the job, and part of the thrill, in the

---

[15]Developed by Bryan Lee, Chris Wills, and Thomas Zhen in the author's course "Embodiment, Tangible Interaction and Games," Auckland University of Technology, NZ, Semester 1, 2011.

Fig. 14 *Tekken Torture Tournament* (*images* Eddo Stern, Mark Allen 2001)

Fig. 15 Brian Jones dies at 27 years of age (*image* The Guardian 1969)

Guitarist dies in pool

Brian Jones, 27-year-old ex-guitarist of the Rolling Stones, died early yesterday after a midnight bathe in the swimming pool at his fifteenth-century farmhouse, at Hartfield, Sussex.

With a Swedish friend, Anna Vohlin, aged 20, and a builder, Mr Frank Thorogood, who was modernising the house, Mr Jones went for a bathe after complaining of the heat. He was found later at the bottom of the pool.

circus, for instance. Also, rock stars live in the fast lane, and many fall "victim to a life-style designed for early death" (Hobsbawm 1995, 324): sex and drugs and rock n' roll! Robert Johnson, Jimi Hendrix, Janis Joplin, Buddy Holly of the Crickets, Jim Morrison of the Doors, Brian Jones of the Rolling Stones (Fig. 15), Bon Scott of AC/DC, Freddie Mercury of Queen, Eazy-E of N.W.A., Kurt Cobain of Nirvana, and Whitney Houston (to name a few) all died young. The risk is part of their performance it appears, and they are compensated for it. It is tolerated or even expected that they have "hilltop houses [and are] driving fifteen cars"; some may have "a bathroom [they] can play baseball in and a king size tub big enough" for eleven people, "a big black jet with a bedroom in it" as well as the "front door key to the Playboy mansion," and they may "date a centerfold" (Nickelback 2005). They are "players" (Ice-T 1991).

Players need security to play, and they find it in a special place that ordinary life cannot touch. Nobody needs to be afraid where everything is play, as Max Frisch says of the theatre. The magic circle of play is conceptually limited to itself and not interested in any effects beyond itself. Nothing that happens within play has any bearing without. Of course, play might stop when it connects too much to the seriousness of the ordinary world, for example, includes dangers for the well-being, social relationships, or material possessions of the players. *Russian Roulette* connects pure chance, a considerable wager of the players, and effects of play in everyday life, and is therefore not acceptable for many people. But even this is an individual and subjective assessment of the players.

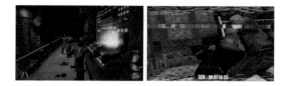

Fig. 16 *Stubbs the Zombie* (demo version) and *Manhunt*

It appears that all games are violent to some degree. Is sport "violence with rules",[16] as is sometimes claimed? Is violence an essential part of play then? Play has performative aspects, which cater to the player's desire to show, to impress, to please, and also to shock. This can be fascinating and add to play's attraction. The graphic violence in many digital games is also a provocation, and intended and used as a distinction of the players from ordinary people and normal life. It allows players to defy socially accepted norms and standards openly, and to reject the (perceived or actual) intervention of laws regulating media content.

Content is not specific to particular media (see Seeßlen and Rost 1984, 24–26). Media's content appears to be largely exchangeable between them and independent of them. McLuhan (2002), among others, points out the distinction between the interaction with media is their content. (Digital) games are not the only medium with (arguably) violent content, but books, comics, and movies are full of it as well (cf. Bolter and Gromala 2003, 98), which has been lamented by worried parties before. Conflict is not different in games and other media, and "conflict is the goad of interaction between characters" (Crawford 2004). Heckhausen introduces the notion of discrepancies [*Diskrepanzen*] to describe the starting situation of play (Heckhausen 1973 in Retter 2003, 48). To equate games with violence appears not to be helpful, and the occurrence of violence cannot be seen as a criterion of differentiation between media.

There is the danger of players perceiving the bodily consequences of playing, for example, with the *PainStation* as unrelated to the game. In the very violent game *Stubbs the Zombie* (2004) (Fig. 16, left) the violence is embedded in the (abstract) mechanism of play, whereas in *Manhunt* (2003) (Fig. 16, right) it appears as an unnecessary and dispensable (graphics) effect. In the former case, players can accept the (excessive) violence as part of play's abstract logic; in the latter it might appear as a disturbing (or appealing) embellishment. With poor integration, players feel rejected from the game, and reject the violence that hinders their play.

---

[16] At the time when young men still individually had to fight their way out of compulsory military service in Germany (see Wallraff 1992), it was common knowledge that doing a competitive sport such as soccer as a member of a club was officially considered to contradict the peaceful attitude of the person and constituted a clear threat to the success of the conscientious objection. Playing badminton was presumably safe.

4.6   Games as Propaganda

Some games have obvious connections into ordinary life. A particular subset of games is intended to function as propaganda for the military of different countries, and to help recruitment. These games challenge a number of characteristics: the purposelessness of play, because they attempt to influence people and openly follow a (political) agenda beyond play; the distinction of play from the everyday world because objects, places, situations, and the like from ordinary life are integrated into play, and it is attempted to maintain this connection between play objects and everyday world objects; and the appearance of play because players might realize and reflect on the meaning of the specific (and not arbitrary) play actions outside play.

America's Army (2002, and many versions after that) is a first-person shooter conceived, paid for, and distributed by the US Army.[17] "Produced with brilliant graphics and the most advanced commercial game engine available (the *Unreal* game engine) at a cost of around $8 million, the game is a first-person multiplayer combat simulation" (Lenoir and Lowood 2012, 36). But players are not expected to pay to play *America's Army*, at least not in cash: "[I]t is free to play online, courtesy of a publicly funded, multi-million-dollar investment by the U. S. Department of Defense" (Dyer-Witheford and de Peuter 2009, xiii).

Many games are made and marketed as commercial products to make money. In general, and in most cases, players presume no agenda beyond this goal (e.g., political). Things are different with *America's Army*. "The aim of this taxpayer-funded project is to generate Army recruits" (Hodes and Ruby-Sachs 2002). It was made as a propaganda vehicle, for "nonstop Army cheerleading" (ibid.). It is openly discussed as a recruitment tool (e.g., by Chris Chambers, former Army major and the deputy director of development for *America's Army* qtd. in Turse 2003), and quite a successful tool, too: The *America's Army* website

> is a major recruitment site for the U. S. Army, one that reportedly has a higher success rate
> in attracting enlistments than any other method. The ... [game] is for the world's undisputed
> armed superpower a serious public-relations device targeted at a generation of game players
> and intended to solve the crisis of a military struggling to meet its intake for the fatal front
> lines of the war on terror (Dyer-Witheford and de Peuter 2009, xiii)

Although not directly asking people to do something (i.e., to join the army), it *primes* them, aiming to get the possibility of joining the (US) army into the player's "consideration set" (Timothy Maude, Army's deputy chief of personnel qtd. in Hodes and Ruby-Sachs 2002).

The game tries to influence people by establishing and maintaining a link between actions in a game and actions in normal life. This is different from most other games that maintain a clear (conceptual) division between play actions and other actions; in some games this is arguably done half-heartedly, such as shooting robots or aliens instead of people (as in *Space Invaders*). But with this game (and one might call it a game), people might be scared or worried, because it attempts to make people do

---

[17]For more on *America's Army* see, for instance, Nieborg (2009).

things in the real world with which a lot of people are uncomfortable. It is a matter of life and death: "Consider ... that the virtualities of *America's Army* cycle into the actualities of combat via the Web link to the U. S. Army home page" (Dyer-Witheford and de Peuter 2009, xix). This might be too much for play.[18]

There are a number of other games in addition to *America's Army* that are produced with similar intentions.

> In 2001 Syria's Afkar Media published Under Ash, in which players take on the role of Palestinians fighting off an Israeli assault; they followed this in 2005 with Under Siege. ... In 2003, developers linked to Hezbollah entered the market with a Special Force series, a set of PC war games set in Lebanon. ... In 2007, Iran's Association of Islamic Unions of Students released Special Operation 85: Hostage Rescue; a first-person-shooter game wherein players aim to free two Iranian nuclear scientists kidnapped by the US (Rayner 2012)

In England,

> the British army launched their online game Start Thinking Soldier in 2009, to drive interest among 16- to 24-year-olds. Then in May last year [2011], China's People's Liberation Army unveiled Glorious Revolution, a Call of Duty-style game for both military and domestic markets. (ibid.)

An overlap of technology between play and other media can be spotted, for example, in computer hardware, software, and interface devices (see, e.g., Seeßlen and Rost 1984, 29, Schmitt 2004, 6). Manovich (2000, 191) observes that "increasingly the same metaphors and interfaces are used at work and at home, for business and for entertainment. For instance, the user navigates through a virtual space both to work and to play, whether analyzing scientific data or killing enemies in Quake."

There are very particular and tangible overlaps between computer technology, games, and the military. Manovich (ibid.:276) notes that the "same interfaces [are used] in flight and military simulators, in computer games modeled after these simulators, and in the actual controls of planes and other vehicles." But it is not only the game interface devices such as flight sticks that are modeled *after* military devices. The exchange works in both directions: "These days [2009] the Commander of a Challenger 2 tank ... communicates with his crew (driver, gunner and loader) by ... [using] a handset almost identical to the Xbox 360 game console handset." This is done to take advantage of its ergonomic design and its familiarity "to the incoming recruits" (The Long Dog 2012). US Marines use commercial games (*Doom 2*, *Falcon 4.0*) to train; and, in turn, the companies working for the US Army on combat simulations release these (with small changes) as commercial games (*Spearhead 2*; see Lenoir and Lowood 2012, 27–33). *Steel Beasts* "is perhaps the world's most successful tank-training simulator"; it was "[d]eveloped by eSim, an American firm with ex-US and European army personnel." According to Nils Hinrichsen, eSim's marketing director, *Steel Beasts* is intended to be "a computer game that [is] both entertaining and educational ... [and to] offer a bit of 'trigger time,' but with accurate procedures and ballistics" (Rayner 2012).

---

[18]Joseph DeLappe's *dead_in_iraq* campaign (see Crogan 2011, 111) establishes a very direct link between an ongoing multiplayer match in *America's Army* and reality.

This is not a recent development. "For decades the military has been using video-game technology. ... Every branch of the US armed forces and many, many police departments are using retooled video games to train their personnel" (Nina Hunte-mann qtd. in Rayner 2012).

Are games such as *America's Army* turning children into soldiers and teaching them to kill? Here it is argued that this happens not more than in *Halma*. Games such as *America's Army* are intended as propaganda. Players might realize that or not. They might work, just as ads on TV. Or they might not work (Huntemann, Crump in Rayner 2012), and players might just play them as they would play any other first-person shooter. Games belong to their players. When they play, they take ownership, regardless of the plans of others. They (re)interpret things, assign new or different meanings, strip things of everyday meanings, draw things into play, and use them for their own purposes. "However powerful the logistical impulse to preemptive control of experience realized in the game design, players open up the possibilities for spe-cific, idiosyncratic adoptions of its entertainment playtime" (Crogan 2011, 174). The meaning of play is the meaning created by players. This is not limited to a particular game genre or to digital games. It applies as well when playing soccer with an old can, and with jackets as goal posts in a car park.[19] This is the magic circle of play, and it is created by the players. The circle exists from the moment the game starts and vanishes when it ends. It is the "special place of meaning" (Salen and Zimmerman 2004, 366), the sacred ground of play. Nobody except the players can enter or touch it (See the chapter, "Makin' Cake-Provocation, Self-Confrontation, and the Opacity of Play" in this volume.).

Players can differentiate between play and nonplay. Otherwise not only educa-tional games would influence and drill people, but also all other kinds of games with violent and questionable content would turn people into monsters. In this regard, play is a medium like other media. There is no direct link between media and life, action, and meaning: "[T]he interpretation of any action's significance is only weakly determined by the action as such" (Suchman 1987, 119). Play activities are highly symbolic and abstract, and only to be understood in the context of a game, by play-ers.[20]

4.7 Pervasive/Mobile Gaming and Live-Action Role Playing

Many games in the past and present are not played on courts, pitches, or boards, but anywhere at any time on which their players agree.

---

[19]Winter (1995) discusses how horror film fans appropriate movies in ways not intended (or imagined) by the producers: "TV is above all a medium that demands a creatively participant response" (McLuhan 2002, 367–368; see also Jenkins 1992).

[20]Therefore, in many sports, there are specialists commenting and translating what is going on for spectators.

These games challenge the characteristics of play's clear distinction from ordinary life, because these games seem to be overlapping or mixed with it, and play's order, for the same reason.

A relatively recent trend appears to be citywide participatory festivals, for instance, the annual *Come Out & Play Festival*[21] held annually in New York or San Francisco[22] (since 2006), and Edinburgh's *New Year Games*[23] (2012). In these events people are "running and sneaking and folding and throwing and counting and hopping" (Hide & Seek Productions Ltd 2013) around the city centers, playing street games such as zombie tag, life-sized *Pong*, and various kinds of mazes. Such events have the character of a spectacle similar to the carnival (e.g. in Cologne), where everyday life is very visibly but temporarily put out of order.

Live-action role-playing (LARP) games are a popular phenomenon and are played in the everyday space of ordinary life. LARPs require a very robust idea of play, thus game-external events challenge but do not destroy play.

There are other games that are played pervasively at odd times and places. The *Tamagotchi* hit the (Western) world quite unexpectedly in the late 1990s; but there are more classic examples. The *Bilboquet* is a French toy. A ball is tied to a short stick; the ball has a small hole; the player throws the ball up and tries to catch it on the stick. It was extremely popular and played by people of all walks of life and of all ages from around the sixteenth to the eighteenth century. Card games have always been played during classes at school, and so on. *Ebay* is also a kind of game where an exciting and competitive second reality is superimposed upon ordinary life.[24]

In *Werewolves*, players take on the roles of werewolf or villager. Both groups attempt to kill each other. It is a discussion-based game (Fig. 17), and everyday levels of authority and trust between people play a significant role in it and are severely challenged at the same time.

A game in which trespassing on the differentiation between play and social interaction is used as a game mechanic is the (German) game *Tot, töter, Geist* (literally, *Dead, deader, ghost*). It is a multiplayer spelling game. One after the other, each player adds a letter to a word that nobody says aloud. At some point, instead of adding another letter, a player challenges the previous player and asks what the word is. Either that player can present a proper word or not. One of the players then loses a life, that is, she goes from alive to *tot*, from *tot* to *töter*, or from *töter* to *Geist*. Players who are ghosts can no longer participate in spelling or challenge players. But they still can (verbally) disturb the game and kill other players by irritating them, for example, by keeping on spelling. It can be tricky to keep track of who is still in the game and who is not. Also, players die when they react to ghosts. This includes

---

[21] www.comeoutandplay.org.

[22] 2007 in Amsterdam.

[23] http://thenewyeargames.com/.

[24] With the increasing participation of shops instead of private persons, and *Buy-it-now* offers instead of regular auctions, *Ebay* seems to be turning into an online warehouse and losing much of its initial playful appeal.

Fig. 17 Students playing
Werewolves: It is day, and
the villagers consult with
each other

answering questions such as, "Anybody want something to drink?" "What time is it?" or "Is it your turn?"

A similar game mechanic is used in *Carré Couper*,[25] a card game for several pairs of players who use secret, game-external signals to communicate with each other.

A game played in public places such as streets is *Ghosting*. Players follow unsuspecting people as long and as closely as possible without being noticed. This is quite edgy behavior that basically asks for (everyday) trouble and conflicts between players and nonplayers who adhere to quite different rules and mindsets.

Planking is a game is which people pose in all kinds of possible and impossible locations as a plank, that is, lying straight on top of things like a block of wood. A photo is taken as proof. Players try to do better than the other players and to "out-plank" each other. *Planking* challenges players to take risks in ordinary life (e.g., planking a billboard) to achieve victory in the game, and rumor has it that occasionally somebody is killed, and players enjoy this element of danger (see, e.g., The Planking Game 2012).[26]

Assassin is played in a group of people over a period of several days, such as when skiing. One person is randomly and covertly assigned to be the assassin, and he tries to kill everybody else by showing them a particular playing card or another symbol at any time. Players can only be assassinated when they are alone; that is, players are safe when in pairs or groups. This usually leads to some peculiar behavior in everyday life.

Crime Plays (2012, Dog Money World, www.crimeplays.com) is a location-based pervasive game played with mobile phones. Players start a (virtual) criminal career by joining "The Syndicate" and are asked to perform certain tasks in everyday life, for example, go to certain locations at specific times, and later, to follow and potentially (virtually) kill other players (Claire Evans of Dog Money World, pers. comm., July 2012). Players are advised that they "still need to pay attention to what is going on around [them]" and to "[u]se good judgment and common sense" when dealing with bystanders (Crime Plays 2013).

---

[25] See https://theoryclass.wordpress.com/2011/04/03/carre-couper/.

[26] The TV series *Southpark* (season 16, episode 3) took up this (fading) meme and invented its own version: *Oh Long Johnsoning*.

The *Middle Eastern Politics Simulation (MEPS)* is an educational tool used to teach undergraduate students at Deakin University, Australia. Hardy and Totman (2012, 190) describe it as "a role-play-based simulation": groups of participants take on the roles of various local and global political players who have an impact or interest in the region, such as politicians, terrorist groups, and international organizations. Participants interact with each other in character through the online exchange of text messages. *MEPS* is typically played by "90 teams and 150–200 students" moderated by two or three teachers for "12 days and with 24 h access" (ibid.:192).

Play needs a strict distinction from the ordinary world to create and maintain tension and ambivalence, but the boundaries of play are conceptual and ideal. Games that overlay everyday life with play, such as LARPs or Assassin demonstrate this.

The games played in the everyday space of ordinary life use everyday spaces and unsuspecting people as part of their play. They are not mixing play and everyday life, though. On the contrary, there is trouble if such an interaction occurs (e.g., in *Planking* or *Ghosting*). But this thrill is made part of play: "In the emerging field of augmented reality and pervasive gaming, the fascination is with the digitally enabled virtualization of real space as resource or affordance for the game's taking place" (Crogan 2011, 23).

It appears naïve to assume play could be or was ever confined to certain places, times, or activities. The magic circle has always been an idea, and the material circumstances have been rather arbitrary. Play's coincidental bleed into everyday life causes accidents and unintended side effects. The *Tamagotchi*, a LARP, and trading cards are identical in that they exist in the middle of ordinary life and have various points of contact and areas of overlap, but are nonetheless strictly divided from it. The order of play is realized in many different ways in different games, and appears as rhythms, cycles, and patterns. The order of a LARP at the city center and the order of a mini golf tournament are different, but both have their specific order.

Many games approach their boundary to ordinary life, play along it and with it. But they only cross it in exceptional circumstances, for instance, through players' high spirits, mischief, or carelessness, and this usually ends play.

5   Conclusion

It is possible to collect and discuss characteristics of play as an activity. But it appears to be of at least equal importance to recognize play as an individual decision, assessment, and position. Play is an expression of the player's mood and attitude. Visible and tangible activities follow from this perspective, invite it, and interact with it. The more players become engaged with play, the more intense it gets, and the less the material expression matters.

Play has always changed its appearance according to what was available in terms of places and materials, and was played with what was available. But the conceptual territory of play is usually intact. Certain games pervert the idea of play, such as *Russian Roulette*. But even that is left to the players to decide.

Pushing the boundaries of play towards everyday life, as it happens in many games, is a problem for play, and it is a question of how these games can be played. Play is a very powerful perspective, and "[a]ny game can at any time wholly run away with the players" (Huizinga 1955, 8), but it is at the same time labile, and "[a]t any moment 'ordinary life' may reassert its rights either by an impact from without, which interrupts the game, or by an offence against the rules, or else from within, by a collapse of the play spirit, a sobering, a disenchantment" (ibid.:21). Players balance play by making sure the characteristics continuously apply to it, and they keep it play, hovering above the material objects and activities that help it to come about. The kick players get out of play is the bigger the closer they can move it towards the edge without falling off. Players decide what they play and what play is (for them). Play appears as an individual and collective perspective of the players.

References

About.com.   (n.d.).   Twister—game.   Retrieved   Mar   7,   2012,   from boardgames.about.com/od/twister/Twister.htm

Alexander, P. (2005, Jan 12). No pain, no game. *The Independent*. Retrieved May 16, 2012, from www.independent.co.uk/news/science/no-pain-no-game-6155108.html.

Bateson, G. (1971). A theory of play and fantasy. In Gregory Bateson (Ed.), *Steps to an Ecology of Mind*. Chicago: University of Chicago Press.

Biskind, P. (1999). *Easy Riders, Raging Bulls. How the Sex 'n' Drugs 'n' Rock 'n' Roll Generation Saved Hollywood* (10th ed.). London: Bloomsbury, paperback.

Bolter, J. D. & Gromala, D. (2003). *Windows and Mirrors. Interaction Design, Digital Art, and the Myth of Transparency*. Cambridge: MIT Press.

Brainy Quote. (2012). Retrieved Feb 2, 2012 from www.brainyquote.com/quotes/authors/d/dennis_rodman.html.

Buytendijk, F. J. J. (1933). *Wesen und Sinn des Spiels. Das Spielen des Menschen und der Tiere als Erscheinungsform der Lebenstriebe*. Berlin: Kurt Wolff.

Cermak-Sassenrath, D. (2010a). Interaktivität als Spiel - Neue Perspektiven auf den Alltag mit dem Computer. Bielefeld: transcript.

Cermak-Sassenrath, D. (2010b). The logic of play in everyday human computer interaction. In S. Günzel, M. Liebe, & D. Mersch (Eds.), Logic and structure of the computer game, (pp. 80–103). Potsdam: Potsdam Univ. Pr.

Cermak-Sassenrath, D. (2013). Makin' cake and the meaning in games. In L.-L. Chen, T. Djajadin-ingrat, L. Feijs, S. Fraser, J. Hu, S. Kyffin & D. Steffen (Eds.), *Design and Semantics of Form and Movement (DeSForM) 2013*(pp. 199–203).

Cermak-Sassenrath, D. (2015). Playful computer interaction. In V. Frissen, S. Lammes, M. de Lange, J. de Mul, & J. Raessens (Eds.), Playful identities. The ludification of digital media cultures (Chap. 4, pp. 93–110). Amsterdam: Amsterdam Univ. Pr.

Chamblanc, F. D. (1828). *Das Kriegsspiel, oder das Schachspiel im Großen*. Wien: Nach einer leicht faßlichen Methode dargestellt.

Château, J. (1946). *Le jeu de l'enfant*. Paris: Vrin.

Christ, K. (2005). *Geschichte der rmischen Republik. Von Augustus bis zu Konstantin* (5th ed.). Munich: C. H. Beck, rev.

Crawford, C. (December 1987). Similarities with other media. *Journal of Computer Game Design*, *1*(5). Retrieved June 25, 2004, from www.erasmatazz.com/library/JCGD_Volume_1/Other_Media.html.

Crime Plays. (n.d). Crime Plays | SAFETY. Retrieved Nov 20, 2013 from www.crimeplays.com/#!safety/c24qc.

Crogan, P. (2011). *Gameplay Mode. War, Simulation, and Technoculture*. Minneapolis: University of Minnesota Press.

Deutschlandfunk. Corso—Kultur nach drei, June 1, 2006. My notes.

Deutschlandfunk. Olympia zwischen Fairplay und Manipulation, Aug 18, 2004. My notes.

Dourish, P. (2001). *Where the Action Is: The Foundations of Embodied Interaction*. Cambridge: MIT Press.

Dyer-Witheford, N., & de Peuter, G. (2009). *Games of Empire. Global Capitalism and Video Games*. Minneapolis: University of Minnesota Press.

Eigen, M., & Winkler, R. (1990). *Das Spiel. Naturgesetze steuern den Zufall*. Munich: Piper.

Fink, E. (1957). *Oase des Glücks. Gedanken zu einer Ontologie des Spiels*. Freiburg: Karl Alber.

Fischer, A. (1925) Psychologie der Arbeit. *Die Arbeitsschule* (pp. 65–76).

Frissen, V., Lammes, S., de Lange, M., de Mul, J., & Raessens, J. (Eds.). (2015). *Playful Identities. The Ludification of Digital Media Cultures*. Amsterdam: Amsterdam University Press.

fritz51338. (2012, June 14). Winners – Go for gold 1988. Retrieved Mar 12, 2014, from www.youtube.com/watch?v=GAnmSvB0Qag.

Game Oven. (n.d). Fingle for iPad. fingleforipad.com.

Hardy, M., & Totman, S. (2012). From dictatorship to democracy: Simulating the politics of the Middle East. In C. Nygaard, N. Courtney, & E. Leigh (Eds.), *Simulations, Games and Role Play in University Education, The Learning in Higher Education (LIHE) Series* (pp. 189–206). Faringdon: Libri Publishing.

Hasbro. (n.d.). Twister for two, three, four or more players (manual). Retrieved Mar 7, 2012, from www.hasbro.com/common/instruct/Twister.PDF.

Heckhausen, H. (1973). Entwurf einer Psychologie des Spielens. In A. Flitner (Ed.), *Das Kinderspiel* (pp. 138–55). Munich: Piper.

Heidfeld, N. (2007, Sep 16). Formula 1 coverage. RTL.

Hide & Seek Productions Ltd. (2013). The new year games : Projects : Hide&seek. Retrieved Nov 26, 2013, from www.hideandseek.net/projects/the-new-year-games.

Hobsbawm, E. (1995). *Age of Extremes. The Short Twentieth Century 1914–1991*. London: Abacus.

Hodes, J., & Ruby-Sachs, E. (2002, Sept 2). 'America's Army' targets youth. *The Nation*. Retrieved Feb 17, 2012, from www.thenation.com/article/americas-army-targets-youth.

Huizinga, J. (1955). *Homo Ludens: A Study of the Play-Element in Culture*. Boston: Beacon Press.

Ice-T. M.V.P.s. LP. (1991). *OG Original Gangster*.

Irving, J. (2006). *Until I Find You*. New York: Ballantine.

Jenkins, H. (1992). *Textual Poachers: Television Fans and Participatory Culture*. New York/London: Routledge.

Juul, J. (2005). *Half-Real: Video Games between Real Rules and Fictional Worlds*. Cambridge: MIT Press.

Keller, P. E. (1998). *Arbeiten und Spielen am Arbeitsplatz: Eine Untersuchung am Beispiel von Software-Entwicklung*. Frankfurt a. M.: Campus. Also: Bremen, University, Dissertation, 1997.

Krämer, S. (1995). Spielerische Interaktion. Überlegungen zu unserem Umgang mit Instrumenten. In Florian Rtzer, editor, *Schöne neue Welten? Auf dem Weg zu einer neuen Spielkultur* (pp. 225–237). Munich: Boer.

Lacroix, M. (2012, Feb 7). Kommentar: Viel Gutes und ein ganz dickes ABER. *Speedweek* (8), 24.

Lenoir, T., & Lowood, H. (n.d.). Theaters of war: The military-entertainment complex. Retreived June 13, 2012, from www.stanford.edu/class/sts145/Library/Lenoir-Lowood_TheatersOfWar.pdf, An updated version (www.stanford.edu/dept/HPST/TimLenoir/Publications/Lenoir-Lowood_TheatersOfWar.pdf (June 16, 2012)) to appear. (In J. Lazardzig, H. Schramm, & L. Schwarte (eds.). (2003). Kunstkammer, Laboratorium, Bühne–Schauplätze des Wissens im 17. *Jahrhundert/Collection, Laboratory, Theater*. Berlin: Walter de Gruyter Publishers.

Magistrix. (n.d.). The Winners – Go for gold lyrics. Retrieved Mar 12, 2014, from www.magistrix.de/lyrics/The%20Winners/Go-For-Gold-132561.html.

Manovich, L. (2000). The Language of New Media. Retrieved Aug 9, 2002, from www.manovich. net/LNM/Manovich.pdf.

Marriott, M. (2001, May 31). Playing games gets serious (and painful). *New York Times*. Retrieved May 16, 2012, from www.nytimes.com/2001/05/31/technology/playing-games-gets-serious-and-painful.html?pagewanted=printŹsrc=pm.

Martin, M., Gavin, J., Cermak-Sassenrath, D., Walker, C., & Kenobi, B. (2013). Shadow showdown: Twister in a digital space. In *Proceedings of the 9th Australasian Conference on Interactive Entertainment (IE2013): Matters of Life and Death*, IE '13, pp. 34:1–34:2. New York: ACM Press.

McCarthy, K. (2001, June 8). Electric shock game controller on the way. *The Register*. Retrieved May 16, 2012, from www.theregister.co.uk/2001/06/08/electric_shock_game_controller.

McLuhan, M. (2002). *Understanding Media. The Extensions of Man*. London: Routledge.

Moss, S. (1959, Jan 24). To really understand the world of men like Mike Hawthorn...their compulsions, fears and exhilarations you have to know their every thought...never before bared so honestly as now. *Daily Express*.

Nickelback. (2005). Rockstar. LP All the right reasons, lyrics from www.lyrics007.com/Nickelback %20Lyrics/Rock%20Star%20Lyrics.html (Feb 16, 2012). Written by Chad Kroeger, Michael Kroeger, Ryan Peake, Daniel Adair.

Nieborg, D. B. (2009). Empower yourself, defend freedom! playing games during times of war. In M. van den Boomen, S. Lammes, A.-S. Lehmann, J. Raessens, & M. T. Schäfer (Eds.), *Digital Material-Tracing New Media in Everyday Life and Technology* (pp. 35–47). Amsterdam: Amsterdam University Press.

Norman,D. A. (2004). *Emotional Design. Why We Love (or Hate) Everyday Things*. New York: Basic Books.

O'Brien, P. (2012, Jan 22). Maradona sticks it up for Messi. *Sport 360°* (464), 3.

Pias, C. (2002). *Computer Spiel Welten*. Munich: Sequenzia. Also: Weimar, University, Dissertation, 2000; Retrieved Oct 28, 2004, from https://e-pub.uni-weimar.de/opus4/frontdoor/deliver/index/docId/35/file/Pias.pdf.

Rayner, A. (2012, Mar 18). Are video games just propaganda and training tools for the military? *The Guardian*. Retrieved Mar 19, 2012, from www.guardian.co.uk/technology/2012/mar/18/video-games-propaganda-tools-military.

Retter, H. (1998) *Einführung in die Pädagogik des Spiels*. Braunschweig: Institut für Allgemeine Pädagogik und Technische Bildung der Technischen Universität Braunschweig, Abteilung Historisch-Systematische Pädagogik, new ed. 2003.

Roch, A. (2009). *Claude E. Shannon: Spielzeug, Leben und die geheime Geschichte seiner Theorie der Information*. Berlin: Gegenstalt.

Roth, P. (2002). *The Dying Animal*. New York: Vintage.

RTL. Verleihung des Laureus 2004. TV, Mar 26, 2005. My notes.

Salen, K., & Zimmerman, E. (2004). *Rules of Play. Game Design Fundamentals*. Cambridge: MIT Press.

Schaller, J. (1861). *Das Spiel und die Spiele*. Weimar: Bhlau.

Scheuerl, H. (1965). *Das Spiel. Untersuchungen über sein Wesen, seine pädagogischen Mglichkeiten und Grenzen* (4th/5th ed.). Weinheim: Julius Beltz.

Schlickmann, D. (2006). June 14). Nowitzki für immer in Dallas. *Sport Bild*, *24*, 68–9.

Schmitt, S. (2004). Feb 2). *Verspielte Forscher. Computer Zeitung*, *6*, 6.

Schnell gesagt! *Motorsport aktuell*, (1–3):28–29, Dec 17, 2013.

Schüll, N. D. (2012). *Addiction by Design. Machine Gambling in Las Vegas*. Princeton: Princeton University Press.

Scott, M. (2012). Aug 7). Es bleibt viel Zeit für das Nichtstun. *Speedweek*, *34*, 36–8.

Seeßlen, G., & Rost, C. (1984). *Pac-Man & Co. Die Welt der Computerspiele*. Reinbek bei Hamburg: Rowohlt.

Sniderman, S. (n.d). The life of games. www.gamepuzzles.com/tlog/tlog2.htm.

Suchman, L. A. (1987). *Plans and Situated Actions: The Problem of Human-Machine Communication*. Cambridge: Cambridge University Press.

The Planking Game. (2012). N.n. Retrieved May 1, 2012, from www.theplankinggame.com/rules-planking.

The Long Dog. (n.d.). Usability—a matter of life and death. In The Long Dog (ed.), *Web, Digital and Communications Blather*. Retrieved Mar 19, 2012, from www.thelongdog.co.uk/?p=482.

Turse, N. (2003). Bringing the war home: The new military-industrial-entertainment complex at war and play. Retreived Feb 17, 2012, from www.tomdispatch.com/post/1012.

Ulie, C. (2001, June 7). The latest sensation in gaming. *InformationWeek*. Retrieved May 16, 2012, from www.informationweek.com/news/6507097.

Wallraff, G. (1992). *Mein Tagebuch aus der Bundeswehr. Mit einem Beitrag von Flottillenadmiral Elmar Schmähling und einem Dialog zwischen Günter Wallraff und Jürgen Fuchs*. Cologne: KiWi.

Weber, S. (2004). *Theatricality as Medium*. New York: Fordham University Press.

Wichert, S. (2006). Aug 20). *Der Schlägertyp. Welt am Sonntag, 34*, 74.

Wikipedia. (2012a). Knife game. Retrieved Mar 14, 2012, from https://en.wikipedia.org/wiki/Knife_game.

Wikipedia. (2012b). Twister (game). Retrieved Mar 7, 2012, from https://en.wikipedia.org/wiki/Twister28game29.

Winter, R. (1995). *Der produktive Zuschauer*. Quintessenz, MMV Medizin: Medienaneignung als kultureller und ästhetischer Prozeß. Munich.

WSRA, Inc. (2012a). Truth or Dare online: How to play Truth or Dare. Retrieved May 7, 2012, from www.tordol.com/gameRulesHelp.htm.

WSRA, Inc. (2012b). Truth or Dare online: Preferences. Retrieved May 7, 2012, from www.tordol.com.

Author Biography

Daniel Cermak-Sassenrath is Associate Professor at the IT University of Copenhagen (ITU), and member of the Center for Computer Games Research (game.itu.dk) and the Pervasive Interaction Technology Lab (PitLab, pitlab.itu.dk). Daniel writes, composes, codes, builds, performs and plays. He is interested in artistic, analytic, explorative, critical and subversive approaches to and practices of play. Discourses he is specifically interested in, are play and materiality, play and learning, and critical play. More information is available at www.dace.de.

Index

Printed in the United States
By Bookmasters